Field Notes from the Northern Forest

Field Notes *from the* Northern Forest

Curt Stager

Illustrated by Anne E. Lacy

S̲U
Syracuse University Press

This book is published with the assistance of a grant from the John Ben Snow Foundation.

The paper used in this publication meets the minimum requirements of American National Standard for Information Sciences—Permanence of Paper for Printed Library Materials, ANSI Z39.48-1984. ∞™

Library of Congress Cataloging-in-Publication Data
Stager, Curt.
    Field notes from the Northern Forest/ Curt Stager : illustrated
by Anne Lacy. — 1st ed.
        p. cm.
    Based upon episodes from a series of weekly radio programs
presented by the author.
    Includes bibliographical references (p.   ).
    ISBN 0-8156-0513-7 (alk. paper)
    1. Natural history—New England.   2. Natural history—Maritime
Provinces.   3. Forest ecology—New England.   4. Forest ecology—
Maritime Provinces.   I. Title.
QH104.5.N4S735   1998
508.74—dc21   97-31275

Manufactured in the United States of America

To Maurice, and to Susan

Curt Stager obtained his Ph.D. in zoology and geology from Duke University in 1985 and has taught biology and guided student research at Paul Smith's College in the northern Adirondacks since 1987. His research interests (ranging from aquatic ecology to global climate change) have led him to East and West Africa, the South Pacific, Sweden, and much of eastern North America. His articles have appeared in *Science*, *National Geographic*, *Wild Earth*, and other periodicals, and he has cohosted an internationally broadcast weekly natural science program with colleagues at North Country Public Radio since 1990. He also composes, records, and performs on guitar and banjo.

Contents

Autumn

Winter

Acknowledgments

Special thanks to Maurice Kenny, who helped to launch this project; to Susan Grimm, who supported me in the long process of bringing it to fruition; to Ken Brown, Martha Foley, and Lamar Bliss, who first created *Field Notes* with me in 1990; and to our listeners. Thanks also to the staff of North Country Public Radio (especially Ellen Rocco); to the scientists and friends who freely contributed their insights and expertise; to the libraries of Paul Smith's College, St. Lawrence University, and Cornell University; to Anne E. Lacy for gracing this book with her wonderful artwork; and to editor Nicole Catgenova, whose timely efforts got these ideas into print.

Finally, I gratefully acknowledge the support and helpful editorial comments of Dr. Ed Ketchledge, who skillfully reviewed and improved this manuscript.

Introduction

Field Notes from the Northern Forest is a collection of short essays about animals, plants, and physical phenomena that are commonly encountered in the forests of the northeastern United States and eastern Canada. It is called "Field Notes" because the essays are based upon episodes from a series of weekly radio programs formerly of that name (now called *Natural Selections*), which I have presented on radio stations across the continent with my colleagues at North Country Public Radio since 1990. I have limited my scope to the "Northern Forest" because that is where most of our listeners live, because it is the region with which I am most familiar, and because I want to focus on a single bioregion for the sake of depth and relevance to those most likely to purchase this book.

First, a bit about the radio program. It consists of four-to-five minute broadcasts in which my cohost, Martha Foley, and I improvise informal conversations about subjects in the natural world that have been suggested to us by listeners, by our friends and families, and by our own natural curiosity. In these episodes, my technical background as biologist and educator plays off against

Martha's perspective as journalist and interested lay person, resulting in a scientifically rich yet refreshingly spontaneous and nonthreatening discourse that enjoys wide popularity among listeners of all ages. It has been my goal to keep that personable atmosphere in these essays, while adding more depth than is possible in a brief broadcast.

As for the term "Northern Forest," I use it to draw attention to a part of the world that gets relatively little media coverage as compared to rain forests and other exotic ecosystems but that is no less complex, interesting, and threatened by overdevelopment and pollution. The Great Northern Forest (as it has been dubbed, with apologies to residents of forests in other northern latitudes!) extends from the Tug Hill Plateau in the western Adirondacks to the coast of Maine and the Canadian Maritimes. It is a patchwork of conifer, hardwood, and mixed forests in which a large number of humans have lived and worked for centuries. At times, human activities have threatened the woodlands with extermination, as when much of the land was logged off in the late nineteenth century. At other times, including much of this century, the woods have moved back onto lands abandoned by people as economic conditions changed.

Today, the Northern Forest faces a renewed onset of development, but this time there is a move afoot to curb unplanned and poorly executed projects and other destructive practices, while fully acknowledging the place of *Homo sapiens* in the landscape. It is an exciting yet increasingly confusing and frustrating time to live in this region, as new regulations are developed, implemented, and revised in the face of local concerns. I hope that in presenting information about these woods and their nonhuman inhabitants, I can contribute something positive to the growth process by helping to encourage more appreciation of what this place is all about.

This is not a field guide; there are plenty of excellent, well-illus-trated books of that sort available. The aim of the book is to take you deeper into the lives of and the processes at work within some of the most common inhabitants of the Northern Forest. This is more than a simple rehashing of old information that can be found in many pot-boiler nature books. I have interviewed experts in various fields and have dug into research results available in the scientific literature to include newly discovered aspects of nonhu-man life in (and under) the forest. Much of this information deals with chemical signals by which animals and plants communicate with each other, defend themselves, and maintain their bodies, and has yet to reach a wide audience among nonscientists. I hope that you will enjoy updating your store of knowledge about these or-ganisms in this manner, and it may even change the way you think about them entirely, as it has for me.

There is little rhyme or reason as to which subjects I have cho-sen to write about here, other than their loose associations with the four seasons. I have simply taken my cues from what has been going on around me here in the Adirondacks over the last few years. The same is true of the contents of each essay; there is much left unsaid in these pages. In some cases, a lack of coverage on a given topic reflects nothing more than my own distraction by another subject. In others, it is because information on a par-ticular aspect was unavailable in my library searches or was sim-ply not yet known to science. In fact, the more that I dig into these topics, the more surprised I become by how little we know about the life cycles, perceptions, physiology, and behaviors of even the most common living things around us. Just try finding someone to tell you what those snowfleas are doing out on the snow in midwinter and you will soon see what I mean!

There are a few other features of this book that I would like to point out here. At the end of the book is a glossary of the organisms and scientific names discussed in each chapter. I have attempted to translate each name into one of several possible English versions and have provided the root words and their translations. All of the word roots are taken from E. C. Jaeger's *Source-book of Biological Names and Terms,* and from D. J. Borror's *Dictionary of Word Roots and Combining Forms,* but the actual translations are my own interpretations and not necessarily those used by others. If you find yourself intrigued by these words and images, try to come up with your own version of an English name; at least, the information should give you a feel for where this tongue-twisting nomenclature comes from. I have also included a complete bibliography of books and periodical articles from which all material other than my own personal observations was taken, in case you would like to read further or to check my sources.

Above all, I want you to have fun with this book. Ideally, after reading each of these essays, you will spend some time outside in your own favorite meadow, lawn, or patch of woods, looking with new appreciation for the things that I have described and using that information as a springboard for your own investigations. The world is an infinite source of wonder and of mystery, and even the most familiar, seemingly tame habitat becomes a treasure trove of surprises when you look beneath the surface, delving into things on different size scales and time scales and probing the world from sensory perspectives other than those to which you are normally accustomed. One of my favorite activities in the woods is to look for individual variations on general biological themes presented in the literature, especially when local goings-on differ from what is thought to be the norm. Does moss really grow best on the north side of a tree? When do the woodpeckers in my back yard start their courtship drumming? If these pages stimulate some of this spirit of questioning and exploration in you, then they will have served their intended purpose.

Before I turn you loose, I would like to put in a shameless plug for our little radio program, for those of you not familiar with it. If you enjoy learning about nature enough to buy this book, you might also enjoy hearing shorter versions of similar things each week on your local public or commercial radio station. *Natural Selections* is broadcast by satellite free of charge to any radio station that cares to pick it up. Ask your local station to tune us in by contacting the station manager at North Country Public Radio, St. Lawrence University, Canton, New York, 13617.

Thanks for listening.

Spring

Ground Bees

Some time in late April, the snow mantle recedes from the south-facing hillside below the small cottage I rent on the forest-cloaked campus of Paul Smith's College. The first warm days of the new spring coax heat waves from the tawny wisps of last year's grass on the sandy slope that descends to the main campus roadway. Along the moist edges of dying snow-drifts, now filthy with sunflower seed husks from my winter bird-feeder, snow-mold spins a delicate mist of cobwebs among the dead yarrow stalks and the compressed skeletons of beech leaves. Male red-winged blackbirds screech from the spiked crowns of silent firs, waiting for their preferred cattail perches to sprout along the shores of Lower Saint Regis Lake, which lies sparkling at the far end of the campus quadrangle below my own hillside perch.

April is a good time to watch this slope closely, and not only for spurts of green growth, though I have come to savor the stag-gered appearance of each species of grass and herb as individual notes in an annual song cycle of botanical rebirth, flowering, and

3

death. I watch most attentively for the appearance of little heaps of loose soil, thousands of them, each about the size of a silver dollar, each surrounding a pencil-thick hole. Above these sandpiles, the warming air soon rustles and blurs with movement, as a widely dispersed swarm of insects patrols the lowest trembling layer of air. A soft drone rises from the warming earth. The bees are back.

When I first heard that sound several years ago, my reaction was purely instinctive: the hair on the back of my neck prickled as my body anticipated a stinging attack. Now that I know these beings well, though, I no longer fear them. I have spent many hours among such bees and have never known one to lose its temper with me, even when I hold them in my open hands. These are not honey bees *(Apis mellifera)*, the kinds that people brought over from Europe years ago to make honey for human consumption and that may sting en masse to protect their communal hives. Rather, these spring bees are true natives and, like many humans, they prefer private property to turf held in common. They are a bit smaller than honey bees, with pale yellow stripes on their black abdomens and with patches of golden insulating fuzz on their heads and paper-winged thoraxes.

When I first noticed the sand heaps on my spring lawn several years ago, it was the low-frequency hum that froze me in place. My heart leaped as I realized that I must have stepped on an underground nest. All I knew then about bees was that they could sting—and that stomping on their nests was a punishable offense. Moments later, I linked the sound to the dark specks darting about my feet, and my heart rate jumped another notch. But the attack never came, and I knelt slowly to be sure that these were indeed bees.

They were, but not like any bees I had ever known. They paid me little heed, even as I blocked their wandering flight paths. One landed on a grass stem beside my booted foot, crawled to earth, and entered one of the sand-hill tunnels. No others followed. A moment later, a single bee emerged and launched back into the air.

From that moment hence I have called them "ground bees" (as opposed to bees that nest in aboveground hives), though I later learned that they have other names as well. Most writers call them "solitary bees," because they dig their own individual burrows in loose sandy soil. But that name is misleading. The bees on my lawn are anything but solitary, for though they keep their own dwellings, their homes are packed into every available patch of dry ground. It is like a housing development. On warm days in April and May, the air traffic is positively urban, and the heat waves above the earthen metropolis quiver with the beating of tiny wings.

It was no easy task finding out what others know of ground bees. Most field guides to insects are packed with pictures and with finely printed details, but it is hard to get much out of them beyond names and morphological features. I wanted to know what those bees were doing on my lawn, and the best way I knew to put my own observations into a factual context was to speak directly with someone who already knew the subject well. Later that morning, I tracked down some of the scientific papers listed in the back of one guidebook and began a phone search of authors listed on the title pages.

Dr. Suzanne Batra was one of these, a noted bee specialist with the United States Department of Agriculture in the Baltimore-Washington area. Her secretary told me that she was out of town during the summer field season but that I could reach her at her summer residence in upstate New York. As I dialed, I realized that it was a local number. Thus I learned that one of the world's leading experts on wild bees grew up within a few miles of my cottage and returns home every year to enjoy her critters.

When I arrived at the red frame house that Dr. Batra shares with her husband (a specialist in fungi, also with the U.S.D.A.) in Saranac Lake during the brief Adirondack spring and summer, she brought me straight to her glass-enclosed porch on the sunny back end of the house. Golden light flooded in from a small pocket lawn lined with rainbow ranks of meadow flowers, warming us as we sank into ancient stuffed chairs amid low shelves laden with books and glass-lidded specimen boxes. Batra handed me a box of what must have been a hundred pin-mounted bees of various sizes and shapes, all banded in shades of yellow and black. Each seemed to hover stiffly in place on two sets of transparent, outstretched wings.

"There are lots of wild bees around here," she began, smiling at the understatement. "All of them are solitary. They are very important pollinators of wild and domestic plants, but most people don't even know they exist. The ones you have on your lawn are *Colletes inaequalis*, a type of pollen bee." In the ensuing hours Batra shared her fascination—even love—for wild bees, which somehow seemed completely in keeping with her ability to skewer them on pins.

Pollen bees differ greatly from the seven known species of true honey bees that pack waxy combs with golden sweetness. The pollen bees, all 20,000 known species of them (over 3,500 in North America alone), dig brood chambers in the soil and feed their subterranean young on pollen cakes mixed with mere traces of honey and other substances that together form what Batra calls "bee bread." Even now when I think of honey bee behavior, I think of dances that communicate the locations of food sources to hivemates and envision a dominant queen ruling over swarms of drones and workers; honey bee society is so complex that I have never been able to keep its details straight in my head. By contrast, one of the things I like about ground bees is the comparative simplicity of their lives. No fussing with overstuffed queens, no worker slavery, no loss of individuality. Among ground bees, everybody gets to do just about everything, including reproduction.

Each bee lies dormant in its respective burrow throughout the winter. When the sun warms the spring earth, he or she rouses from torpor and digs to the surface. The sand piles that result from the emergence of thousands of bees make the lawn look like it is being mined by giant ants—which is not all that far off, because ants and bees share the same insect order, Hymenoptera. The order name stems from the Greek words *hymen* (membrane) and *pteron* (wing) and describes the double sets of transparent, delicately veined wings typical of most of its members. Wasps also belong to this membrane-winged group, but they are distinguished from bees (in part) by their largely carnivorous diets.

Batra plucked a specimen from among the dense ranks of mounted bees in the specimen case and handed it to me. I held it gingerly with the pin pedestal between my thumb and forefinger, taking care not to knock off the stiff little wings. It was a bit smaller than the others, hairless, and uniformly brown in color. "This is an insect that causes plenty of grief for your bees," she said. "It's called a cuckoo bee. A female cuckoo bee is a brood parasite who lays her eggs in the nests of other species, like the cuckoo bird does. She sneaks down into a pollen bee burrow when the owner is off foraging, lays an egg on the pollen store, and tries to fly off before she gets caught. I suspect that her small size probably helps her to get in and out of that burrow as quickly as possible. Later, the cuckoo bee grub hatches out and eats the pollen intended for the host's young. In fact, it probably also eats the pollen bee eggs as well."

A ground bee's polyester capsule

With Batra as my mentor, I set out to learn as much as possible about ground bee life. My particular bee metropolis occupies a south-facing slope, the first bit of ground to warm up in spring. Presumably, the bees take advantage of that solar heating to get an early start as soon as the snow melts off and to make the best use of available warmth at a time of year that can be cold and cloudy much of the time. Once free of their underground birthplaces, the bees launch into a frenzy of digging, breeding, and foraging that ends in a mass dying soon after the last tree leafs out.

Clearly, these creatures were zooming about my lawn with much to do in a very limited time. Between late April and early June they had to excavate new brood chambers, provision them, lay eggs in them, and seal them away from the outside world of hungry birds and inclement weather. By mid-June, every adult would die of old age.

During that first spring, I focused on the features that had first caught my attention: the burrows. Sprawled prone on the crunchy grass on a particularly warm dry day and taking care not to block any doorways, I chose one female for close inpection. I could tell that she was a female because she was busy digging a new burrow (only the females do this). She was just starting to work a patch of loose soil beside me, pawing the ground with her front legs like a dog burying a bone. A steady stream of fine sand sprayed from between her back legs until, moments later, she was out of sight. She backed out every thirty seconds or so with

Ground bee with capsule

an armload of debris, dumping it behind her on a growing heap downslope. When I got up to stretch an hour later, she was still at it.

Most of the burrows on my lawn were dug within two or three days. This was easy to document: I roped off a square meter of lawn and marked the dozen burrows inside it with upright twigs. The next day I went out and marked the new burrows that appeared in that same square. There were half a dozen more on day two, and only two more on day three. Extrapolating to the rest of the lawn, I figured that there were at least five thousand mounds all told, each sporting a tidy round entrance hole. When heavy spring rains plugged the entrances with debris, the bees reopened their doorways as soon as the sky cleared and continued digging.

Although one tiny hole in the sand looked pretty much like another to me, the bees did not seem to have any trouble telling which tunnel was which as they came and went. They may memorize landmarks around the burrow, such as pebbles, weeds, or fallen sticks, and the females may also use their antennae to home in on chemical substances with which they have marked their individual tube entrances and which guide them home even in the midst of a dense colony.

I was eager to see what was going on in those tunnels, but could not bring myself to dig any of them up. Although I am a biologist and therefore given professional license to "sacrifice" living things for the pursuit of my investigations, I have always had great difficulty in doing so. There are many compelling rationales for some measure of destructive sampling in field work, and I do not begrudge others for doing so when the cause is truly important. But rather than harden myself against feelings of sympathy for living objects of study in the pursuit of knowledge, I generally prefer to enjoy what the natural world offers me in its own time and manner.

That is not to say that I am above taking advantage of opportunities generated by the activities of others. In late August of that first bee year, I was granted a guilt-free means of probing the underground world of *Colletes inaequalis*. The college administration decided a new water pipe system was needed before winter set in. When I heard the first rumble of the backhoe on the hillside, I spent a brief moment feeling sorry for the bees, then rushed over with trowel and camera in hand. The water line trench tore a ten-by-ten-foot gash through the heart of the colony. As the backhoe noisily chewed its way up the hillside at the far end of the trench, I slid down the sandy wall at the opposite end.

I found no burrows: the bees had apparently backfilled them all. Instead, I found papery brown capsules scattered throughout the upper layers of the soil, mostly within a foot of the surface. They seemed to be made of caramel-colored cellophane, and they resembled medicine capsules in size and shape. Those that were intact were full of a soft, dense substance, but many lay crinkled and empty, presumably left over from earlier years.

I plucked one of the fat capsules from the sand—and nearly dropped it as it buzzed furiously in my hand. There was an adult bee inside, very much alive, and apparently very displeased by the disturbance.

"You can watch the females build those little capsules if you confine them for a while in a sort of ant-farm of plate glass filled with sand," Batra later explained over the telephone. "They dig the chambers out during the day, but they also continue working through the night underground. After nightfall they line the insides of the chambers with a thin film of liquid squeezed out of their mouths, which hardens into a form of polyester."

Polyester? The same polyester that cocktail lounge musicians wear?

"Yup," she replied. "It's a biogenic plastic. Sometimes I call them 'polyester bees' because of it. Not only is it a biogenic plastic; it's both breathable and water-repellent, and is biodegradable. Once the polyester lining hardens, they fill the chamber with a pea-sized ball of pollen and nectar and lay a single egg on the pile. Then they seal the chamber off to form a self-contained polyester capsule, backfill the tunnel, and move on. The egg hatches in a few weeks, and the larva feeds on the pollen ball."

I hung up the phone, amazed. Plastic as a "natural" substance invented by insects. Stepping back outside to ponder the hillside in the light of this new knowledge, I imagined the subterranean seeds of next spring's bee population sleeping beneath the grass. In that dark underworld beneath the probing rootlets and percolating rain, grublike larvae hatched inside their breathable capsules, eating sweetened pollen and metamorphosing into adults, their wings bound close against their backs by tight maternal wrappings. For most of their adult lives they would lie quietly, unaware of the world of illuminated air just above them, until the first fingers of spring warmth coaxed them sunwards.

The following April, I resolved to watch the mass emergence from the beginning and to see what bees do in their first days of life above ground. No longer would this bee metropolis be a random swarming zone to me. I now knew how to distinguish males from females (males are slightly smaller and have slightly longer antennae), and I had a hunch that the males would appear first. This is a common reproductive strategy among migratory birds, such as the screeching redwings; long before the gals arrive on the northern breeding grounds, the guys stake out their territories and sing lustily at each other and at their phantom mates.

Sure enough, when the first darting forms appeared over the spring lawn, I found only males. What were they up to? They certainly were not eating, because nothing was in bloom yet. Many of them buzzed in and out of the dark bank of spruce and fir trees lining the upper border of the slope, sometimes landing momentarily to clamber among the prickly needles, as if searching for something. After several more minutes of following them in their seemingly random patrols, I learned what was on their minds. It was not food, but rather the other thing that both birds and bees do a great deal of in the full glory of spring.

I sat beside one individual as he clung to a withered blade of grass overhanging the entrance of his burrow shelter, absently cleaning his antennae with his forelegs. Without warning, a small dark blur slammed into his back and seized his abdomen in a frenzy of beating wings and flailing legs. A moment later, the dive-bomber let go and dashed off in search of more satisfying targets, and my little neighbor soon resumed his grooming as if nothing had happened. I supposed that even the bees themselves sometimes have trouble telling each other apart, especially in the heat of passion.

A few days later, the first females appeared on the lawn as well, but sex seemed to be the last thing on their minds. They immediately set to work digging new brood burrows and provisioning them with pollen that they ferried in from the newly flowering willow thickets. The amorous males would have none of this, though. In order to start a brood burrow, each female had to work face down in the dirt, leaving her attractive abdomen exposed to hundreds of cruising suitors. I watched in amusement (and a bit of embarrassment) as two Romeos pounced simultaneously on one female who had been trying to excavate a sand pile beside my right foot. Noisy mayhem ensued. One male held her down briefly but the other shoved him aside and mounted her from the rear. There was a desperate bending, thrusting, and pulsing of behinds, as if for stinging, as the female's abdomen filled with sperm. After a full minute of voyeurism, I averted my gaze—only to find another such encounter under way beside my other foot. In fact, the entire hillside was littered with tumbling pairs and with single females who cowered under any available cover. It is a wonder that they got any work done at all.

The only quiet refuge for a female was indoors. Looking carefully into the open sand-rimmed burrows, I found most occupied by fuzzy yellow faces that retreated into the darkness if I came too close. I could not tell for sure that it was only females who sat thus in their doorways, but I figured that that is where I would probably be if I ever found myself in their situation.

A raft of cloud skimmed its cool shadow across the sun-soaked hillside. The bustle of the bee metropolis slowed momentarily, as the air and earth cooled. When brightness drove the shadow eastward a few minutes later, the hillside once again resonated with sound and motion and a new awareness of this place dawned inside me. This lawn is alive, and not just in the botanical sense. There is probably almost as much tunneled air space, pollen, and bee protein in its skin as there is dirt. I winced as I thought of the coming summer session, when students in the college's surveying program would swarm over the sleeping rookery, as they have always done, and pound sharp wooden reference stakes into it.

A week later, I was just finishing lunch when I felt the rumble of heavy machinery through the legs of my dining room chair. The college's road maintenance crews were bulldozing winter windrows of loose dirt away from the roadsides in preparation for graduation day. I ran to the door just in time to see a bucket loader scrape away the soil where, earlier that morning, I had watched one particularly energetic female at work. I rushed down the hill to the road's edge, where I thought her burrow might have been, and cast about for signs of life.

The man at the wheel was mystified. He shut off the engine and leaned out of his cab. "Whatcha looking for?"

"Bees," I shouted over my shoulder, still probing the scarred soil.

He tipped the bill of his hat back and straightened slightly. "BEES?"

The motor went idle. Soon a half dozen maintenance fellows had gathered around me (at a respectful distance), staring and shaking their heads as I stooped and gently lifted a wriggling form from the wreckage. It looked enough like the female I was seeking that I felt a surge of relief. The skin of my hand beneath her belly tingled with a soft buzzing, though her wings lay neatly folded across her back. She seemed to be shivering her internal wing muscles to warm them up before taking flight.

"Will ya look at that," the driver muttered as dozens of confused-looking bees emerged from the scraped surface, all apparently unharmed. Their burrows were deep and the digging had only shaved off the tops of their tunnels. "I thought those things were anthills. Look at all them bees!" His companions cast nervous glances around them at what had once seemed to be an empty lawn. I could almost see the transformation at work behind their eyes. "They don't seem too eager to sting anybody though. What are they up to, anyway?"

The following Monday dawned clear and fresh. I looked up from my breakfast coffee toward the sandy roadside, and smiled. Down on the scraped earth where the bee lawn met the tarmac, a group of college students huddled around a uniformed, shovel-toting maintenance man. They tipped their hats back and leaned closer as he knelt and pointed with his free hand to something down in the dirt.

Small Worlds

People are often disappointed when they come to the Adirondacks in order to see wildlife. They read up on the bears, deer, and otters that live here, and they bring binoculars to spot eagles and rare warblers. They spend a day or two in the woods, see nothing more than red squirrels and chickadees, and go home wondering where all the animals went.

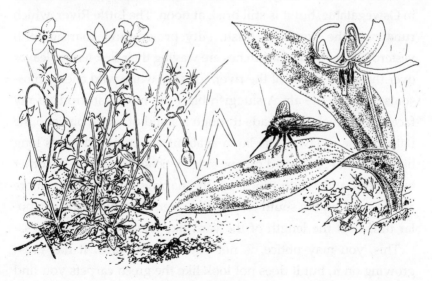

It does not have to be that way. The Adirondacks are crawling with life, but the bulk of that life is not the stuff you see in field guides and on nature videos. Most of it is small, compared to us, and is not furred or feathered. It is diverse, beautiful, savage, and as exotic as any tropical rainforest. Most of it is still virtually unknown to science. And you can find many of the more interesting things going on in very accessible places, such as back yards. In fact, the problem of not finding wildlife often is not the result of some lack in the environment; rather, it lies in the preconceptions of the observer.

All you have to do is to shift your scale of perspective down to that of an insect, and the world transforms before your eyes. Kids do it all the time, when they lie down on their bellies in tall grass and pretend that they are in a jungle. When was the last time you did that? In case it has been a while, let me take you on a tour of the lawn outside the little red cottage where I am writing this chapter.

Let's step outside the screen door into the May sunshine. Onchiota got a dusting of snow last night; we just got chills, here in Oswegatchie, but it is still brisk at noon. The Little River, which runs past the porch, is flush, silty brown, and roaring with yesterday's rain, and the birches are waving their first tender leaves over the rocky banks of the river. Creamy white wild cherry blossoms sweeten the air. A sluggish black fly hovers in front of my face. The gravel drive leads through a stand of white pines to the flat expanse of lawn. The lawn is about one hundred yards long by thirty wide, bordered on one side by Scotch pines, on the other by aspens and white pines, on the narrow north end by more aspens, and on the south end by the river. Cottony wisps of poplar fluff float the length of the yard like enormous snowflakes.

This, you may notice, is not your normal yard. It has grass growing on it, but it does not look like the green carpets you find in the suburbs. It has big brown patches and blotches of dark greenery in it, and blobs of sky-colored bluets *(Houstonia caerulea)* are scattered across it. No fertilized, pesticided monoculture, this. Things actually live here. Lots of things.

Here is one trick to seeing more life outdoors than most people do. In a situation where many things are happening at once, you may do best to approach it without any particular observational goal in mind. Do not go to your lawn only to look for brown bee-flies (family Bombyliidae); just go to your lawn. Let things come to you unbidden, and pay attention to whatever arises. Maybe, sooner or later, you will actually see a bee-fly. But even if you do not, you will still get caught up in whatever is going on at the moment and learn things you would not have learned otherwise.

With that in mind, lie flat and look around. The plants suddenly spring into finer focus. Vibrant green trout lily *(Erythronium americanum)* leaves jut above everything else like mottled garden trowels. Tiny beaks on their tips helped them to pierce through last year's blanket of dead leaves. A clump of bluets looks more green than blue, as their leggy stems and short leaflets crowd the view; the upward-opening blossoms are aimed to draw attention mainly from above. Down in the understory, wild strawberry plants *(Fragaria virginiana)* unfold white petals around central clusters of golden pollen-rich stamens. Clumps of freshly sprouted grasses spring from open spaces like groves of bamboo amid twisted tangles of winter-killed stems and leaves. Dark hedges of squat, bristling haircap mosses *(Polytrichum* species) shoulder out competitors in dense patches and conserve body moisture by reducing exposure to the open air.

Every fragment of ground here is occupied. Where no plants grow, the lumps, cups, sheets, and branches of gray and green lichens gather like coral formations. The only sign of the sandy soil below is a dark, gritty heap of debris piled up by tunneling ants. The undersides of neighboring trout lily leaves are splattered with bits of this brown grit, reminders of what yesterday's downpour must have been like for the inhabitants of this miniature world.

Right away, you can see how different small habitats are. For one thing, fungi (especially those forming the lichens) are much more prominent in small landscapes. If you were an inch tall, there would be many forms of lichen that would brush your waist and some, like the branching reindeer lichens (*Cladina* species), would stand taller than your head. On our human size scale, they are little more than scattered crusts and tufts of pastel color. Fungi are limited to small size scales by the physics of water transport. They lack the fluid-transporting vessels of most plants and must therefore obtain water from their surroundings by slow osmosis and by sponging capillary action. A ten-foot-tall lichen could not draw water up from the ground and into the upper parts of its body fast enough to keep itself alive.

From the prone perspective, this lawn is a veritable forest. Peer across the tops of the trout lilies, and the lawn becomes a vast expanse of jungle. Heat waves, invisible from above, set the air above the leaves atremble. The aspens at the far end of the lawn rise like Himalayas on the horizon, as inaccessible and irrelevant to life down here in the jungle as distant mountains. Bees and blue butterflies hover like tropical birds over the flowering canopy.

Down here, signs of animal life appear under our noses almost immediately. Sunlight glistens on a fine thread of silk stretched between two trout lily leaf tips. A tiny, bead-like red-brown spider, probably freshly hatched, dangles from one loose end, about an inch above the lichens. It must have drifted in on that string parachute and is now trying to decide what to do next. A gentle breath of wind stirs the hair on my arm, so lightly that I barely notice it; to the baby spider, though, it is a hurricane blast. The spider whips back and forth on its thread like a pendulum, swinging the full length of each arc in less than a second. A moment later, it is still again.

Let's think about this. What would it feel like for us to swing on a pendulum that quickly? Imagine hanging from a twenty-foot rope and swinging all the way forward and back in a little less than a second. We would have to travel one full arc at about ninety miles an hour, reverse instantly, and swing another arc in the opposite direction at the same speed. Even Tarzan did not swing that fast.

Life on size scales smaller than ours is fundamentally different. It is just the way that this physical universe works. Think of those scenes in low-budget movies where a ship plows across a stormy sea. You can tell right away that the scene was shot with a small model boat; the waves and splashes just do not look right. That is because water does not behave on tiny size scales the way it does on large scales. On tiny size scales it is more like molasses than water, which is why it does not form convincing spray and roughness in the model scenes.

Small creatures have to be able to move much faster than we do, even taking into account proportional size differences. Fortunately for them, their small size also tends to make them proportionally sturdier and stronger. Take, for example, that little red ant working through the tangled grass stems and lichen formations, probably following an invisible scent trail laid down by a comrade. Think of what it must be like for her (she is a member of the nonbreeding female worker caste) to cover the distance from my right hand to my left. From my perspective, I am placing them on flat ground ten inches apart. To the ant, my hands are not truly on the ground. They are resting on crushed vegetation and dark crawlspaces well above the actual surface of the ground. And the distance between them is much more than ten inches; it is at least twice as far, because the route is a three-dimensional obstacle course. The difference for the ant is like the difference between following a sidewalk and clambering through forest blowdown. Even so, it only takes about a minute for her to cover the distance. That is because small creatures such as ants can move incredibly fast, as a result of their size. See how little time it takes for the ant to decide where to step next and then to trundle from one tilted, unsteady grass blade to another. A split second. That would be like me traversing the length of a fallen tree, branches and all, in the same length of time.

Now press your face down among the flowers and grass stalks, and breathe deeply. It smells like a greenhouse down there. But no such odor was noticeable when you stood erect; your head was immersed in different air currents. In among the plant stems, air travels less freely, even on windy days; the spider swung so violently because it hung high in the lawn canopy. In the deepest recesses, there may be no wind at all. Moisture evaporating from the soil and transpiring through green leaf tissues humidifies the sheltered mini-atmosphere; you can feel the moisture dampen your clothing where you lie upon the living ground. Down in that lush greenhouse, aromatic vapors rise from the rich humus and from the growing things, thickening the confined air.

In the very thinnest layers of air that lie close to the surfaces of leaves and lichens, air is essentially motionless. This still layer is called a "boundary layer," and it forms in fluids as well as in air. Currents slow down from the friction of air molecules against solid objects, and the closer you get to the object itself the slower the flow. That is how dust builds up on the whirling blades of an electric fan. You can see the boundary layer effect at work in the lawn by looking closely at the surfaces of the leaves. In the right light, infinitesimal specks of what seems to be yellow dust glow on the green background; it is pollen, probably from the surrounding pines. No wind will blow that stuff off; only a down-pour could remove it.

I wonder what wind feels like to an insect. Air molecules are proportionally larger for an ant. If the animal is a quarter of an inch long, air molecules are nearly three hundred times larger to it than they are to me. Do they feel different when they tumble into you when you are a quarter-inch long? Perhaps the wind is thicker and heavier for baby spiders than it is for us; all they have to do is to spin a length of silken rope out into it and the viscous, swirling currents lift them aloft.

Water also acts differently when you are small. For an ant, being struck by a raindrop is a much more serious mishap than a simple soaking. As I pointed out earlier, water is a sticky, cohesive liquid that acts like glue on small size scales. If an ant were to stumble into a droplet of water bigger than itself, resting like a crystal ball on the edge of a grass blade, the surface tension of the droplet would tend to drag her in toward the center of the drop. If she managed to struggle free of that watery trap, she would still be covered with a heavy, clinging goo that could not easily be escaped until it evaporated. This is probably why you do not see many insects out in the rain.

Even warm sunlight is different when you are small. Have you ever seen those glass vacuum bulbs with the foil leaves inside that spin when you shine light on them? The impact of the sun-light particles (photons) pushes the foil hard enough to move it in the absence of air friction. An ant probably weighs about as much as one of those foils. Do ants feel sunlight as a forceful impact, as we feel the streaming of water while standing in the shower?

The haircap mosses fold their leaves up like umbrellas when they get too dry, reopening only when rain or dew soaks them. This drying can happen very quickly in small organisms exposed to full sunlight. Lying out in the noon sun, the heat prickles our bare arms. But for the ant and for the baby spider, a short time in that heat can be dangerous. Insects breathe through pores in their abdomens that they can close off temporarily to reduce the evaporative loss of body fluids. Spiders have somewhat similar setups. One of the commonest ways for larger animals to cool off is to sweat or to pant, thus releasing excess heat through the evaporation of body water. But if a tiny ant tried this, it would dry out very quickly.

A cloud arrives overhead, briefly blocking the direct sunlight. For small creatures, the slight shift in temperature directly affects their body temperature and water balance. Physiological mechanisms that worked full bore to prevent cooking must shift gears immediately, but be ready to jump back into full swing as soon as the cloud passes. Precisely how the insects and arachnids in this lawn regulate the delicate trade-off between breathing, body temperature, and fluid retention in the hot sun remains a mystery.

There are many more kinds of animals living down in the small forest than you might think. Spread a sheet of paper out on the ground in a place such as this and see what lands on it. A springtail seems to appear out of nowhere, then disappears just as suddenly. Three miniature gray leaf-hoppers, each the size of a period and probably newborn, briefly occupy one corner, then vanish. A black hunting spider darts out onto the paper, hesitates as if surprised by the large, flat open space, and darts back the way it came. Midges and other tiny flies arrive on iridescent wings; one has red eyes like a fruit fly. Two kinds of ant, one red and one black, meander across the paper, going who knows where. All this during five minutes of watching.

Here is another point to keep in mind. Most animals on this planet are arthropods (having jointed limbs and armorlike cuticles). People talk about insects inheriting the world after we wipe ourselves out in a nuclear Armageddon, but it's already a fait accompli. Insects took over the world hundreds of millions of years ago. If space aliens arrived to sample typical Earth life, they could reasonably report back to home base that the predominant animal life forms here have six legs.

You will never run out of exotic-looking animals to observe if you develop your insect sensitivity. The tamest-looking habitat is a howling wilderness on the small scale. Rhinos, spotted owls, and humans may come and go, but they are so large as to be virtually unnoticed by much of the life on the planet. We are as huge to small things as drifting clouds are to us. When we pass, some of the larger beings in our path may duck for cover and reemerge when we are gone; but most of the events in the area never skip a beat. For quarter-inch-long life forms, the planet is hundreds of times larger than it is for you and me, and the diversity of habitats and species is staggering. We will probably never list and describe all of it, much less understand it.

Interest in insects is often an acquired love, of course; it was for me. I was fortunate to be able to take a graduate course in the subject from an excellent instructor (Dr. Fred Nijhout, who still teaches at Duke University), but you could also rid yourself of gut revulsion by educating yourself with a good set of field guides, a sweep net, and a hand lens. It is well worth the effort, because the wealth of insect diversity and beauty is boundless. Even if you never find a particular creature in a field guide and therefore cannot affix a formal name to it, you have nonetheless met it in the field. Make up your own nickname, if you need a label for it.

With insect sensors turned on full, we can appreciate the sight of a flying brown speck zipping toward us from the far end of the lawn. It zooms up to the nearest patch of nodding bluets, its fingernail-sized body fuzzy with soft, insulating brown fur. Two

wings beat too fast for human sight as it hovers motionless over
a blossom. That blue petal corolla tilted skyward and released
fragrance all morning into the upper air currents just for the sake
of this moment. A stylus of a snout, perhaps a quarter of an inch
long, protrudes from the ball of fur and pokes into the gold-
ringed heart of the bluet, seeking nectar. A momentary thrust, a
hum of wings, and the alert creature darts away as quickly as it
came.

We have finally seen a bee-fly. Built much like a bee and thus
enjoying a measure of protection from predators that can't count
(a real bee would have four wings), this bee-fly is harmless to
humans, fascinating to watch, and elusive enough to be a challenge to find. Learning a great deal about it is also a challenge,
one that could take more than a lifetime to pursue. You will not
find much about it in a field guide, and because it does not carry
human diseases, damage crops, or produce marketable products,
you can bet that few if any scientists have been funded to pay it
much attention.

You are often on your own when you venture into the small
wilderness, as alone in your seeking as an Amazon explorer. You
will frequently have to pose and to address your own questions.
What species was it? It is not in my field guides, so I will just call
it "brown bee-fly" (*Fly-us bee-us*, perhaps?). Where and how did
it spend last night, to survive the near frost? Where did it just go?
How long will it live, and how will it die? What is it like to be a
bee-fly? Mysteries, mysteries all around us.

Woodpeckers

*Hairy
woodpecker*

I heard my first woodpecker of the new year pounding on a tree last March. They had been coming to my suet station all winter, but I had not heard them drilling away yet. That is because woodpeckers peck most loudly on wood when they want to be heard by other woodpeckers, and they only care about other wood-peckers when they want to establish a territory or to find a mate. So, you mainly hear woodpeckers pecking during their courting and mating and nesting seasons. It being late March, I figured that things were right on schedule.

You can go out into the woods at most any time of year and see solitary woodpeckers busily inspecting rotten trees, prying away at dead bark and chipping into soft dead wood with their bills.

They do not make much noise when they do that. They are just hunting for hidden insects, and they do not want to draw undesirable attention from anybody with a taste for bird flesh while they hunt. The pileated woodpeckers (*Dryocopus pileatus*), which look like prehistoric feathered monsters with their huge bills, black and white bodies, and flaming scarlet head-crests, often pry their reluctant prey out of fallen logs rotting on the ground as well as from standing trees. They are so big and powerful that they can just rip bark and wood apart to expose dinner. One of their favored treats is hibernating black carpenter ants, which they excavate from the wood of balsam fir trees in winter.

Smaller species, such as the hairy woodpecker (*Dendrocopos villosus*), sometimes take advantage of the work of their larger relatives, rushing in to glean insect leftovers from the bare wood. Small woodpeckers also keep an eye on their chickadee neighbors and follow them when they find food. When chickadees and woodpeckers show up together at your winter feeder, chances are good that the woodpeckers are tag-alongs. But even the smallest woodpeckers are very efficient at finding food in what can seem to us like barren, featureless wood. Some researchers suspect that hairy woodpeckers choose the best spot to drill by pressing their bills against tree trunks and feeling for the minute vibrations of munching grubs.

Some kinds of woodpecker do not dig all of their food out of tree trunks. Yellow-shafted flickers (*Colaptes auratus*), those medium-sized brown birds with the white rump patches (which, I am told, you are now supposed to call "northern flickers"), like to rustle around on the leaf-littered forest floor looking for ants and such to eat. Flickers have to migrate south in the winter because their food supply gets buried by snow.

Then there are the acorn woodpeckers (*Melanerpes formicivorus*) that live in the American Southwest and get their name from the habit of digging holes in the trunks of oak and pine trees and stuffing them full of acorns. The current record for overall size of stash is about fifty thousand acorns in a single tree. There may be more to this practice than simple food storage; some individuals are said to store pebbles instead of acorns! Or maybe they are just a bit wacky. One report describes a bird that spent an entire season dropping hundreds of acorns through a knothole in a cabin wall.

Drilling a hole into a tree is only one part of the problem of extracting tasty treats from solid wood. You also have to be able to reach into the hole and to pluck reluctant insects out of it. Woodpeckers solve this problem nicely with the help of a mouth harpoon. That is what their tongues are like; pointed, stiff, and barbed. Sapsuckers (*Sphyrapicus* species) have little bristles instead of barbs on the ends of their tongues; the bristles trap air spaces that act like capillary sponges with which the birds "suck" sap. Actually, it's not really sucking, but who wants to call them "sapsoakers"?

And woodpecker tongues can be quite long. The green woodpecker *(Picus viridis)* sports a tongue four times as long as its bill. The thing is so long that its muscular roots start at the base of the bill, encircle the back of the skull (under the skin), fuse on top of the head, and end up sheathed in the right nostril. Next time you get a chance to look at a roadkilled woodpecker, you might consider checking out its tongue. I did it once as a student at Bowdoin College, after hearing about the green woodpecker. Ignoring the honking of outraged motorists, I pulled over for a dead hairy woodpecker lying on the shoulder of Route 1 in southern Maine. I remember feeling the sharp tip of the tongue protruding from the gaping bill and being glad that I was not a soft-bodied grub awaiting its lethal thrust. Back in my dorm room, I peeled back the feathery skin and laid bare the straplike tongue, pale against the deep red muscles of the head. I tried to imagine what it must be like to feel your whole head spring to life every time you stick your tongue out.

When you hear a woodpecker from a distance, it sounds like it is beating triple-time on a drum. That is totally different from the regular pecking after food, because the whole aim of it is to make as much noise as possible. It is called "drumming," instead of "pecking," for that reason. While most birds are busy screaming their heads off at each other in spring, woodpeckers are drumming their heads off instead.

Even the kind of tree that a woodpecker uses for drumming is carefully chosen. The bird usually looks for trees that are not rotten, because even though rotten wood may be good for finding grubs and beetles in, it is about as resonant as a wet towel. Also, think of how fast a woodpecker's drumroll is supposed to be; if it tried that on a dead tree, its bill would probably be as effective on the soft, punky wood of the trunk as an ice pick on cheese. Drumming-trees are usually hard and dense, so when a woodpecker slams its bill into it the tree resonates like a bell and throws sound out in all directions. I would bet that, if you took the time to look, you would find that there are only so many super-resonant trees in your woods that woodpeckers bother to drum on and that the same ones probably get used year after year. People have told me that they have even heard woodpeckers making a racket on metal objects such as chimney pipes, lamp posts, and trash cans. A friend of mine from nearby Rainbow Lake once complained about a particularly annoying bird that kept him from sleeping late on weekends by pounding on the metal roof of his house.

It would be interesting to see whether woodpeckers share drumming trees or hog the best ones for themselves. I am sure that it depends on the kind of woodpecker, and perhaps even the individual bird, although verifying this by personal observation would be hampered by the problem of trying to identify individuals.

Few animals use musical instruments other than those built into their own bodies. I remember hearing chimpanzees drum on resonant tree trunks in the rainforests of West Africa, and I know quite a few of my closer relatives who can lay down a mean rhythm track as well. But as far as bird-music goes, woodpeckers are the only ones I know of that use rhythm more than melody to send a message.

One of the most common kinds of woodpecker around my part of the Adirondacks—and for that matter in forests all over North America—is the hairy woodpecker. "Hairies" look much like the equally widespread, five-to-six-inch-long downy woodpeckers *(Dendrocopos pubescens)*, with black-and-white checkered patterns on their wings, but they are about half again as big as downies. The broad distribution of hairies means that you can find them in wildly different climates, ranging from the wintery woods of northern Canada to the rainforests of Central America. I have always wanted to travel the length of that range, mainly just for fun but also to keep an eye out for the different versions of hairies that live at different latitudes. There is a well established biological principle, called "Bergmann's Rule," according to which animals in colder climates often tend to be larger than compatriots living in warmer climes. This presumably helps them to conserve body heat during winter, because larger bodies have proportionally less surface area from which to lose heat.

Downy woodpecker

Speaking of body heat conservation in winter, I recently read an article by Bernd Heinrich in which he reported seeing downy and hairy woodpeckers excavating holes in tree trunks in November, several months before the nesting season. Heinrich suggested that the birds were preparing shelters in which to spend frigid winter nights. Having a small enclosed space like that can be a great help to a cold bird, because it retains body heat that would otherwise be lost to the surrounding air. If you keep careful watch in your own woods, you might be able to tell if your local woodpeckers do this as well.

I have read elsewhere that hairies have two kinds of territory in which they spend most of their time. One is big, about seven acres or so, and the hairies sometimes share it with other woodpeckers during the year. But when the breeding season rolls around (midwinter to early spring), look out. Each couple sets up a much smaller territory—maybe one hundred feet across—centered around their nest tree, and they keep other woodpeckers out. The sound of their territorial drumming serves as an auditory *No Trespassing* sign to potential intruders.

It is a fair bit of work for a pair of seven-inch birds to peck out a nest cavity from a living tree trunk (most often relatively soft aspen or poplar, but sometimes maple or oak) in two or three weeks, so it should not be surprising that the birds are possessive about their homesteads. You can often locate nesting sites by looking for pale wood chips scattered around the base of a tree, because the birds are not very careful about hiding evidence of their work. Both male and female hairies incubate the white, inch-long eggs, with the males usually taking the night shift. The busy parents forage within a fairly short distance from the nest for most of their daily allottment of wood-boring beetle larvae, ants, moth larvae and pupae, spiders, and the occasional berry or seed.

While we are on the subject of bird food, I ought to mention something about suet. It is great for attracting woodpeckers to your feeder, because fat (from the bodies of plump, juicy insects) makes up the bulk of their diet. Trappers wintering in the north woods have reported hairies picking bits of fat from freshly stretched pelts. But if you like woodpeckers you should be sure to stop putting the stuff out as soon as the weather warms up. Beef suet, like any fat, melts when it is heated. Imagine yourself as a bird, poking your bill into a mass of buttery warm suet, and you can figure out what can happen to a woodpecker. It gets filthy wet with the stuff. This mats the feathers of its head, which means that they cannot keep the cold and rain out like they should, and the rancid mess becomes a breeding ground for harmful microbes. The bird may lose its facial feathers entirely. Better to close up shop early than to risk dehairing your hairies.

Among hairies, protection of food resources is perhaps the main reason for their aggressive territoriality; overcrowding of nest sites means less food for everyone. But there is also a risk of unscrupulous bird villains hijacking newly excavated cavities rather than digging their own. I once read about an English sparrow that was seen plucking baby hairies out of a nest and dropping them into a nearby river, presumably to evict them before taking over the nest. The author of that report never found out the real reason, in part because he shot the sparrow before it finished its nefarious work, but similar behavior is common among non-native sparrows and cowbirds.

Hairy woodpeckers can live fifteen years or more, and you would think that they would stick with the same nest cavity year after year, because they put so much energy into excavating it. But they do not seem to do this very often, much to the delight of house-hunting starlings, sparrows, and mice.

If you have hairies in your woods, and you are also kind of nosy, you have a good chance of figuring out who drums where. There is usually a male and a female in each territory during the mating season (midwinter to spring), and they are pretty easy to tell apart from a distance. The male is the one with the bright red patch on the back of its head. Both of them drum, but they usually do it on different trees. With a bit of snooping, you might find out whether each sex prefers a specific kind of drumming-tree. You may even find mated pairs drumming to each other on occasion, in call-and-response fashion. Individual hairies also have their own unique black and white patterns on their wings; if you have a good pair of binoculars and you know where the local drumming-trees are, you could try to find out if different individual hairies share them.

If your vision is as bad as mine, though, it would be much easier and just as interesting to find out if different species of woodpeckers use the same drumming-trees. Around my place, we have tiny downies, slightly larger hairies, and enormous pileated woodpeckers. All of them drum, but the speed of their drumming decreases with size, so you can recognize them even when they are out of sight. Downy-drumming is about as fast as the clattering sound of a twig that has gotten stuck in the spokes of your bicycle while you are moving at a moderate pace. Hairy-drumming sounds like you have slowed down a bit to check out that pesky twig in your spokes. Pileated drumming is something altogether different; it sounds like someone attacking the side of a barn with an ice axe.

It is amazing to think that these birds choose the hardest of trees to knock their noses against, but woodpeckers are built to beat the living daylights out of solid objects. Unlike most splay-footed birds, typical woodpeckers use two clawed toes pointed forward and two pointed back to clamp them firmly to rough bark for forceful striking. Stiff tail feathers also press tight against the trunk, further enhancing the leverage of powerful neck muscles.

This all means that woodpecker heads have to withstand violent collisions better than wood does. They have hinged bills and internally padded skulls to absorb some of the shock, and extra-hard tips on their bills reduce the risk of breaking their beaks off. Even so, it must take dedication to make music with your face like that, especially in the chill of late winter and early spring. Although it probably would not be of interest to the woodpeckers, they have one appreciative listener in me. When I hear those first drum rolls resonating in the woods, I know it will not be long before the avian spring chorus begins.

Underground Connections

One of my favorite times to walk in the woods is early May, just after the last (cross your fingers) snow melts and just before the black-fly hordes awaken. Waiting for the spring wildflowers to appear, my mind probes beneath the surface of the ground, and I try to imagine what is going on down there in the dark, moist realm of the soil. Even after the forest leafs out and the

Indian pipes

spring beauties and trout lilies give way to summer's wood sorrels and Indian pipes, my imagination still finds plenty at which to wonder in the underground realm. Flowers, trees, and mushrooms draw people's attention through their aboveground features; the flicker of green leaves, the nodding of petals and branch tips, the puffs of spores and of scents. But we experience only a part of the forest in this way. There is yet another world beneath our feet.

It is particularly easy to sense this in early spring, when empty branches spread like roots against the sky. Sometimes I can sense the whole picture by thinking of shore-hugging trees reflected on the still surface of a lake, where the shoreline divides mirror images of trunks and branches above and below. If the soil suddenly vanished, the whole forest would look something like that, with a thin membrane of leaf litter spread between branching limbs and branching roots. Most of the action in the plant world goes on out of sight of humans, hidden by slower time scales and by the opaque film of dead leaves and humus.

What really got me thinking about the underground was preparing lectures for a college course that I was asked to teach on forest soils. In the process, I became fascinated by something I had never heard of before. The subject was *mycorrhizae* (my-ko-*rize*-ee), and it permanently changed the way I look at forests.

The direct translation of the name itself means "fungus roots" (from the Greek words *mykes* and *rhiza*), a rather succinct explanation of what it is all about. It is a mutually beneficial fusion of subterranean fungi with plant roots. Mycorrhizae were discovered only in the last century, and most scientists and foresters have only paid close attention to them in the last few decades, but their importance to the survival and structure of forest ecosystems makes our past ignorance of them rather embarrassing. Nearly all forest plants have helpful mycorrhizal fungi attached to their roots, and many cannot live without them. In fact, some seeds do not even bother sprouting until they are penetrated by the vital fungal threads.

Perhaps I should give you a bit of background on fungi before going further. Once classified as "degenerate plants," fungi are now assigned to their own separate kingdom. Members of the kingdom Fungi include mushrooms, toadstools, puffballs, molds, yeasts, and mildews, of which there are many species. What most fungi have in common is the ability to reproduce with spores and a threadlike composition (as opposed to the rather blocky or blobby cell structures of most living things). Under a microscope, the sliced mushrooms gracing your salad look like slabs of pressed cotton, each minute fiber the rough equivalent of a cell.

A single fungal fiber is called a *hypha*, derived from the Greek word for "web" *(hyphe)*. A hypha looks like a tendril of spider webbing, only much more delicate; hyphae are often no thicker than one thousandth the diameter of a human hair. This is what grows out of a spore when it settles on a suitably damp, organic substance such as a fallen log—or the sole of an athlete's foot when he or she steps into a public shower. Hyphae absorb water and nutrition from living or once-living matter. In the former instance, fungi are a cause of disease; in the latter, they are the agents of decay, recycling life-sustaining nutrients in ecosystems. By themselves or buried deep within the tissues of dead or dying plants, the pale cobweb strands of individual hyphae are practically invisible to the naked eye. But they are everywhere, penetrating tree bark, saturating the duff of the forest floor, and spreading gossamer webworks into the darkest depths of the soil itself.

An underground network of hyphae, or the mass of hyphal threads that makes up the body of a mushroom, is collectively called a *mycelium* (my-*see*-lee-um; again, based on the Greek word for fungus, *mykes*). The complete mycelium of a mushroom is much larger than the fleshy umbrella you find in the woods or on your lawn; that is just a temporary spore-dispersal tower built from spare hyphae, some water, and a great deal of empty air space. The main body of the fungus is much more diffuse. This is the principle behind fairy rings, those mysterious-looking circlets of mushrooms that you sometimes encounter on lawns and in woodlands. The fairy ring mushrooms mark the leading edge of an expanding network of hyphae that fills the soil in the seemingly empty heart of the circle. Although most mushrooms are more randomly dispersed than that, most of them are just as interconnected below ground.

A mycelium can be many years old, far older than the short-lived fruiting bodies that dispense spores and quickly wither. And some are unbelievably huge. One of the largest organisms yet discovered is a mycelium of the mushroom *Armillaria bulbosa* living in a Michigan forest. The mycelium, born of a single spore more than a thousand years ago, weighs about one hundred tons and covers thirty acres. Every *A. bulbosa* mushroom in that patch of woods is just a lump on the back of that same enormous mycelium.

Now let's get back to mycorrhizae. As a mycelium spreads, it eventually encounters the roots of plants growing within its borders. Swarms of hyphae surround or even penetrate the microscopic root hairs that plants use to draw water and nutrition from the soil. It is this fusion of fungal and plant tissues that we are calling mycorrhizae.

Like a global communications network that bridges continents and cultures, the mycorrhizal tangle forms trade routes between organisms. Sugars sent down from a plant's leaves may be used by the root cells for growth energy, or they may be passed on to the hungry mycelium. Unlike fungi that digest starches, such as the woody cellulose that stiffens the cell walls of plants, many mycorrhizal fungi seem unable to survive on their own. They need ready-made sugars, the nutritional molecules that other fungi obtain by breaking apart cellulose and other plant starches; one excellent source of such sugars is plant sap. The host plants also provide vitamins unavailable elsewhere, and sometimes they even release fungal growth substances that encourage the connections. The mycelium, in turn, acts like a vast extension of the plant's natural root system, passing water and vital nutrients along to the plant in the process of feeding itself. Several distinct fungi may even connect and combine their efforts as well, further extending the mycorrhizal system into the surrounding soil. Plant roots that team up with such fungal helpers gather more nutrition from the soil than those that do not and are apparently favored over the long run by natural selection.

In the cases of many species of willow, beech, and birch trees, the very shapes of the roots change when the correct species of mycorrhizal fungi attach to them. The hyphae inject plant growth substances into the roots that stimulate the development of stubby, forked mycorrhizal rootlets that resemble branched coral formations.

This dependence of plants on specific mycorrhizae can also work to the advantage of their competitors. Pine litter, for example, releases chemicals that specifically inhibit the species of mycorrhizae necessary for the growth of many herbs and hardwood trees. In like manner, some free-standing, ground-dwelling lichens release chemicals into the soil that inhibit conifer mycorrhizae, thus reducing the survival of neighboring conifer seedlings that might otherwise crowd them out.

Take the white pines (*Pinus strobus*) that cling to the sandy ridge near my home, for example. The glacial sand is very nutrient-poor, and water percolates through its porous fabric so quickly that trees living on it risk death by drought. Fortunately for them, several dozen kinds of mushroom mycelia can form mycorrhizal networks with white pines and help trap enough soil-water to keep them alive. Sometimes pines do not even bother to produce water-soaking root hairs at all, because the hyphae take their place so effectively. Fungal threads sheathing the rootlets of one pine extend far and wide in the porous soil and can link up with the roots of other individual pines at the same time. Substances produced in one tree may thus be transferred through the mycelium into neighboring pines. It would be interesting to try to find out whether trees use these mycorrhizal networks as chemical communication highways. In any case, it is clear that anything that might harm the fungi in that sandy ridge would threaten the lives of those pines as well.

It is not unthinkable that something could damage the fungi in mycorrhizae. Before foresters knew better, it was a common practice to treat fungal diseases in commercial forests by spraying the whole stand with fungicide. This often made the trees even sicker.

Mycorrhizal fungi typically colonize new forest stands by extending hyphae up to the soil surface to form mushrooms that scatter spores on the wind. But animals sometimes play a role in this system as well. Some experimental forest management schemes have called for the mass poisoning of red squirrels (*Tamiasciurus hudsonicus*) in conifer stands slated for replanting. The idea was that because squirrels eat conifer seeds, they must inhibit forest regeneration. Surprisingly, the newly planted conifer seedlings withered in squirrel-free zones. Careful ecological sleuthing finally illuminated the problem. Nocturnal flying squirrels (*Glaucomys sabrinus*) were also being eliminated along with the reds in the extermination programs. Both kinds of squirrels love to eat mushrooms; flying squirrels even make night raids on the food hoards of sleeping red squirrels in order to steal their stored fungi. As a result, squirrel droppings are loaded with fungal spores and are an important dispersal mechanism for mycorrhizal fungi in the Northern Forest. Without the squirrels, seedlings may have difficulty establishing their necessary fungal connections.

Awareness of other forms of underground connections lends even greater depth to woods-walking. For example, many plants are themselves directly interconnected, because they grow out of underground runners or root systems. Quaking aspens (*Populus tremuloides*) are a classic example of this principle of connectedness. Aspens form entire groves by the asexual production of shoots from a single root network. The shoots produce full-sized trees that can grow forty to fifty feet in height and that share moisture and nutrients with each other. This system presumably benefits stems that occupy poorer soil than their relatives do. Aspens grow quickly and can rapidly colonize areas exposed by fire and by other disturbances. In addition to sharing root-borne fluids, individual aspens gain an additional edge in growth by maximizing their capacity to capture sunlight for photosynthesis. Aspen bark has a greenish cast to it, because its inner layers are packed with green chlorophyll; this allows the interconnected clones (genetically identical individuals) to continue to trap sunlight after their leaves have fallen. Even the characteristic fluttering of quaking aspen leaves lets extra light through to the lower branches, thus helping the whole canopy to photosynthesize more effectively.

Sassafras *(Sassafras albidum)*, bigtooth aspen *(Populus grandidentata)*, certain mosses, and many other plants form linked associations in much the same way that aspens do. Strawberry *(Fragaria virginiana)* plants even do it above ground. But aspens remain the world champions of plant interconnectedness. Not long after the mushroom-that-ate-Michigan story hit the presses, investigators from the University of Colorado nominated an aspen grove in Utah as The World's Largest Organism. They named it "Pando" (Latin for "I Spread") and pointed out that their enormous aspen clone covered about 100 acres, weighed over 5,000 tons, and contained 47,000 stems. The last I heard of this rather silly scientific competition, Pando the aspen clone was still the champion.

In case you would like to measure the aspen clone nearest you for comparison to Pando, here are a few methods for telling which trees belong to the same root system. The average angle between branches and the main trunk is thought to be a genetic trait, and the precise angle supposedly varies somewhat between clones (although I must admit that I have never been able to tell them apart this way). Different clones also produce leaves at slightly different times in spring; that is a good time to mark the members of a stand with colored flagging, for future reference. In autumn, you can go back to your marked stands and look for variations in color as the leaves change from green to a particular shade of yellow or gold. The leaves of each clonal cluster should have a unique but uniform color. The whole clone normally begins to change color all at once, but it should be a bit out of phase with its neighbors.

This all means that your favorite patch of trees may actually be a single enormous organism. Include the subterranean mycorrhizae in your mind's eye view, and the entire collection of branches, stems, roots, and soil fungi in your woods becomes one gigantic breathing, eating, growing being. It makes me wonder what constitutes an individual. Am I killing an aspen when I cut it down? Or am I just trimming the hair on the back of a huge organism?

For that matter, what about me? My body consists of swarms of living, breeding cells that arose together from a single fertilized egg cell in much the same way that a great *Armillaria* mycelium grows out of a single spore. Technically speaking, that makes me a colony of microscopic clones.

Another interesting concept involves clones that associate so loosely that they spread out over the face of the land, allowing us to experience them as being separate individuals. Aphids are an example: for part of the year, female aphids breed asexually, producing genetically identical female offspring that also breed in like manner. The main difference between them and my cells or *Armillaria* hyphae is that these insects are not physically connected to each other. Biologist Dan Janzen has suggested a mind-numbing solution to this dilemma: why not treat a lineage of aphids like a single superorganism that spreads itself out so thinly that predators cannot find and destroy the whole being?

Like the trees we have just discussed, many wildflowers have secret lives underground as well. Pink lady's slipper orchids *(Cypripedium acaule)*, for example, have roots that branch so little and are so devoid of water-absorbing root hairs that they look like spaghetti noodles. Lady's slippers survive only because mycorrhizal fungi feed them. The seed of a lady's slipper is so small that there is barely enough room for the tiny embryo inside it, much less the starchy food supply that most seeds hold in store to give the sproutling a head start in life. But that is not a problem for this particular seed, because it is prepared to lie dormant in the soil until soil fungi find it. Sometimes the intruding hyphae kill and consume the seed; that may be one reason why orchids produce many seeds at a time. But, for reasons unknown, the fungus may also choose not to attack, instead providing a vital link to the mycorrhizal underworld. The young plant may spend two years or more hidden underground, living solely on fungal supply lines, before it rears up into the open air.

Lady's slippers, of course, are not the only wildflowers in the Northern Forest that depend on mycorrhizae for survival. The roots of Jack-in-the-pulpit *(Arisaema triphyllum)*, trout lily *(Erythronium americanum)*, wild ginger *(Asarum canadense)*, false Solomon's seal *(Smilacina racemosa)*, and *Trillium* species are equipped with unique passageways and protuberances in their roots that encourage fungal hyphae to connect with them.

To me, the prize for most intricately interconnected wildflower goes to the Indian pipe *(Monotropa uniflora)*. You may have seen clusters of Indian pipes in your woods in summer, although I would not say that they are common. Indian pipes are insect-pollinated flowering plants, but they look much like fungi because they are white and fleshy-looking rather than green. The stems and scaly leaves of Indian pipes do not have the green chlorophyll pigments most plants use for photosynthesis. Why? Because they do not have to.

Until recently, many botanists assumed that Indian pipes get their food directly from decaying organic debris in the forest floor, in much the same way that fungi do. Others figured that they were parasites on tree roots. It was not until 1960 that a lone researcher figured out that Indian pipes are connected to mycorrhizal fungi just as most plants are and that the fungi sustain them by extracting nutrition from soil humus. The most common kinds of fungi involved in this relationship include those that produce spongy-looking *Boletus* mushrooms. If you were to drive a sheet of metal a couple of feet down into the soil between your cluster of Indian pipes and the nearest tree, the Indian pipes would likely wither and die. If you were to squirt a chemical tracer into an Indian pipe's sap vessels, you might eventually find it in the pines whispering above your head. Or maybe in the *Trillium* flowers nodding beside your foot, or in the royal ferns *(Osmunda regalis)* over there in the shadows. It turns out that Indian pipes tap into the same mycorrhizal network that their green plant neighbors are using.

Body fluids flow every which way in that underground switchboard. Personal boundaries dissolve under such conditions, and the whole concept of independent existence in forest plants comes into question.

Personally, I find this cooperation between organisms to be rather comforting, especially in light of the habit many people have of describing the natural world solely in terms of struggle and competition. The more we learn about life on this planet, the more such cooperative interactions we find. Why fight over resources when everybody can benefit by working together? Quite a valuable lesson to be learned from hidden roots and hyphae in the woods.

Trillium with **Boletus** *mushroom*

Why Do Bugs Bite?

Being a Northern Forest dweller for most of my life, I have often had cause to wonder when and where the world's worst bug season really is. Many places claim to have the bloodthirstiest black flies or the most enormous mosquitoes, but these claims are often hard to substantiate because people rarely support their claims with hard data, such as "bites-per-minute," that can be used for objective comparison. Consider, for instance, a scientific study of mosquito abundance conducted in the Canadian Arctic in which researchers disrobed and dutifully recorded nine thousand bites-per-minute, enough to drain off half of your body fluids in two hours.

While this sort of scare mongering can be useful when you are trying to keep tourists out of a favorite patch of woods, I find it much more interesting to contemplate why some insects bite you in the first place and what their individual lives must be like. It takes determination to set aside an instinctive revulsion for swarming six-legged creatures that suck human blood and to consider things from the viewpoint of the biters. But it is worth trying, if only to develop some sense of a purpose behind the suffering they cause.

Actually, before we go any further, we need to get something straight. We are not really talking about bugs here. We are not really talking about bugs here. Neither mosquitoes nor black flies are true bugs, in the scientific sense. They are flies (order Diptera), which you may have already guessed if you speak enough Spanish to notice that "mosquito" can be translated to read "little fly." Most true bugs (order Hemiptera) rarely bother humans. If one bites you it is probably because you picked it up or scared it, like I did once when I rolled up my shirtsleeve and jammed my right hand deep into a bucket of freshly collected green pond slime in front of a group of impressionable urban high school students. My intent was to show them that there is nothing to be afraid of in pond slime, but I succeeded only in having my index finger painfully lanced by a giant water bug (*Lethocerus*). The kids loved it.

But unless you muck around in bug-infested slime on a regular basis, it is most often flies—deer flies, midges, and gnats—that go for you aggressively. In the Northern Forest, the two most notable types are black flies and mosquitoes. Both are quite small by our standards, but their bites pack a whallop all out of proportion to their size, and in spring and early summer they appear in such numbers that they can drive even the hardiest folk indoors. Here in the Adirondacks we even tell time by these insects; "bug season" comes right after "mud season," usually some time in mid-May. In good years, there is a brief window of opportunity between the two seasons when the ground is dry, the air is warm, fresh, and inviting, and the trails and lakes are relatively free of seasonal visitors. This is the glorious week or two in which I try to set all other obligations aside and to spend as much time outdoors as possible.

Last spring I stood knee deep and sleeveless in a trout stream near Meacham Lake, casting over and over even though the fish were not biting, just to feel the sun and water on my winter-bleached skin. But much of the sweetness of such moments in the Northern Forest actually lies in the knowledge of what is soon to follow. Sensing the inevitable, I paused to kick up a grapefruit-sized cobble from the stream bottom and held it close to examine its slick surface, knowing from past experience what I would find. There, like a dozen ticking time bombs, clung the dark threadlike larvae of black flies.

Adult black flies resemble squat, hunchbacked versions of house flies, only shrunken down to about a quarter of an inch or less in length. They are usually the first biters to appear in force with the coming of spring. When not feeding, pregnant black flies drop their eggs into or near moving water, where they sink quickly to the bottom and adhere to solid objects such as rocks and logs. Some species gather by the hundreds in communal egg-laying swarms, laying thousands of eggs in the same spot.

After hatching, a typical eyelash-sized larva holds tightly to submerged surfaces by a hooked foot and silken anchor strands, filtering food particles out of the current with a pair of fluffy "mouth-brushes" atop its head. If the current is very swift, black-fly larvae position themselves on the sheltered downstream ends of rocks (or soda cans, grocery carts, or any other fixed object) where they are less likely to be cast adrift and consumed by trout or by predatory aquatic insects. As it nears maturity on the stream bed, each larva builds a tiny tent-shaped shelter in which it pupates and then metamorphoses into a winged adult. The last challenge in this submerged stage of life is to traverse the overlying stream currents to reach open

Black-fly larva

air. The black fly's solution is both elegant and effective. It produces a bubble of gas drawn from the surrounding water, in which it can sit like an aquanaut inside a minisub. Under its own buoyancy the bubble-encased fly rises to the surface, where the floating globe bursts and the tiny creature lifts above the water on dry wings.

It is only such fully grown flies that bite, and of these it is only the females that draw blood, for reasons that I will explain shortly. If you ever should magnify the face of a black fly under a microscope you will see the tools of her trade. From beneath glistening compound eyes and various bristly, segmented appendages protrudes a set of scissorlike jaws. When these chew into your skin, they bisect enough tiny vessels to leak a blood meal into the open wound. That is why black-fly bites often bleed.

One of the greatest difficulties I have in relating to biting flies is in trying to conceive of them as individuals. There are good reasons for this, of course. It is hard to tell the individuals apart because they are so small and because—being insects with stiff exoskeletons instead of pliable, expressive faces—they do not have the kinds of facial features we look for in recognizing each other. Their very numbers make them plural rather than singular entities in our minds. All of these factors make it difficult for even the most dedicated scientists to learn much about them. Just imagine trying to radio-collar or band a mosquito as you would an elk or a migratory bird, and you soon come to realize that we may never learn as much as we might wish about what it is like to be a biting fly.

This should not keep us from making the effort, however. With a little imagination and some field observations, one can often get a general idea of what the life of an insect may be like. Let's try it with a mosquito.

Black-fly pupa

As I write this, it is early June in the Adirondacks. During the daylight hours, the black-fly season is still in full swing, and gardeners wear head nets and gloves and tuck their pants cuffs into their socks. When the sun set a few hours ago, the black flies retreated to the sheltering leaf litter and foliage of the forest, but outside my office window warm-blooded mammals such as myself still find no respite from winged attack. Scattered randomly over the outer surface of my window are eight light brown bodies, each about the size and shape of a grain of rice. I move close to focus on one individual, examining the slender underbelly. Six long legs as fine as human hairs splay out from the front half of the body, with the front two angled forward, the back two rearward, and the middle pair outstretched. A pair of translucent wings lie folded neatly atop the back, extending to the tip of the abdomen.

I know that this is a female because she has very short antennae; male mosquitoes have long, brushy antennae, which they use to trace faint pheromonal scent trails to potential mates. I encountered a cloud of such males earlier at dusk, as I walked along the lakeshore beside the college library. I heard them well before I saw them. A thin, high-pitched drone seemed to come from the air itself, and I stopped and stared upwards, my eyes focusing at different depths to probe the apparent emptiness. They materialized out of the gloom then, perhaps a hundred male mosquitoes in a diffuse whirling column, their wings whining a signal to all interested females that this was the place to come for a good time. They seemed to follow me as a group, until I walked faster than the swarm seemed willing to drift. They were not really after me per se but were more likely using me as a landmark. Or, perhaps, as bait to help draw in potential mates, the insect suitor's equivalent of a walking box of chocolates. Perhaps the female now resting

Adult black fly

on the vertical glass face of my window is fresh in from that lakeshore trysting site.

I have no way of knowing, without capturing her for microscopic study, to which one of the 170-odd North American species this mosquito belongs, though I would guess she is a member of the genus *Aedes*. But I can deduce something about her past. Most mosquito mothers in these parts lay their eggs in dense floating rafts on the surfaces of still, stagnant pools and puddles such as those in the marshy area across from the building where I write. Many choose their egg-laying sites according to water chemistry and color. Pools of water in which rotten wood or weeds have been soaked attract more egg-layers than does pure water, and dark-bottomed containers or those filled with ink-stained water are more attractive than are clear-looking pools. This makes sense if you consider that dark, stagnant, organic waters are more likely to have abundant microbial nutriment for developing young than would clean, sterile water. This particular mosquito may have spent the winter underwater as a twitching, grublike larva, or perhaps she overwintered in egg form and hatched only a few weeks ago in some nearby pool, marsh, or stump-hole. As an aquatic larva she hunted floating microbes, but as an adult she has fed only on plant sap, fruit fluids, and nectar. Some mosquitoes are important pollinators; perhaps she carries flecks of golden flower dust on the fine sensory hairs spiking her legs.

But tonight she is after something different. Her reproductive tract is freshly loaded with sperm, and hundreds of fertilized eggs are growing inside her abdomen. Each egg is a living, growing cluster of cells draining her maternal tissues of protein and minerals. She must replace those losses or die, and mere sugar water is not rich enough in the necessary compounds. Only blood will do. This is why biting flies bite, and why it is only the females that do so.

Naturally, we tend to notice only those insects who take notice of us, and thus live unaware of the countless other lives underway beyond the limits of our consciousness. Many species of mosquito do not bite humans at all. *Aedes hendersoni,* for example, seeks her prey only in the leafy canopies of eastern forests, specializing in the blood of squirrels. Members of the genus *Culex* are more likely to pester birds than mammals; one northern species accepts nothing but the blood of loons.

When I step outside this building later tonight, I will be attacked within seconds. It will not be an easy chase, because even my usual walking pace is faster than mosquitoes can fly, but wherever I go, the volatile essences evaporating from the surface of my skin and clothing will leave a swirling trail of scent and warm carbon dioxide exhaust.

Try to imagine how it might feel to be a mosquito lying in wait as the ultrasound radar of hunting bats resounds all around you. The sensory bristles sprouting from your exoskeleton probably tremble with those echoes, tugging at hair-triggered nerves inside your body. You launch into the deadly airspace in spite of the bats, and fly towards a hulking shape. A blast of warm, odiferous mammalian breath buffets you, and you strain forward, hovering to scan for a safe landing spot. You lunge for a tuft of hair in a sparsely carpeted spot, seize it with one leg, then latch on with the other five as well. Leaning downwards and tipping your tail end up, you press the tip of your slender needlelike mouthparts against the wrinkled, oily skin floor of this hair forest. It penetrates like a sliver piercing soft dough. You continue to push, and your sword probiscus bends like a spring as it slides forward about half a millimeter beneath the surface. You work the serrated inner units of your elongated jaws back and forth like saw blades, probing and slicing in search of widely scattered capillaries. Your saliva creeps into the wound, keeping the blood flow strong with its anticoagulent proteins. If you happen to hit a nerve, you will be crushed instantly.

Blood fills the saw-torn epidermal chamber, and twin suction pumps in your head draw it quickly into your gullet as your abdomen swells like a red balloon. If you are lucky, you withdraw safely after about a minute, well before the tissues around your feeding station swell and itch in telltale allergic reaction to your salivary proteins. You lift heavily into the air and retreat to the nearest vegetation to rest. There you will gradually expel most of the blood's bulky water in a single clear droplet through the tip of your abdomen, and your hunting will be over. All that will be left for you to accomplish in your few remaining days of life will be to locate a suitable water body on which to lay your eggs.

There is a particularly unpleasant aspect of insect biting that we tend to forget about in North America: parasitic disease. Living animal bodies are excellent nutrient-rich habitats for microscopic organisms, and some parasites not only dwell within insect bodies but also migrate to the bodies of yet other species as they grow and metamorphose. Because of this, biting flies are potential carriers of illnesses; West African black flies, for example, can carry onchocerciasis (river blindness), and tropical mosquitoes can spread malaria, yellow fever, and filariasis.

Fortunately for those of us living in high latitudes, most of the really nasty parasitic diseases are now largely limited to the equatorial regions, although other temperate-zone animals suffer greatly from insect-borne parasites. Dogs catch heartworm from mosquito bites, for instance. But consider the insect-related health risks some people cope with daily, as a matter of course. More than two hundred million people, most of them living in the tropics, are suffering from malaria at this very moment. In New York, you call in sick and people assume that you have the flu. Call in sick in equatorial Africa, and people are more likely to assume that your malaria is acting up. We used to catch malaria here in North America as well; New York City experienced major malaria epidemics at the turn of this century. Filling and draining wetlands where *Anopheles* mosquitoes bred helped put an end to this scourge for most North Americans. But even today you occasionally hear about someone in the United States acquiring an encephalitis virus from a mosquito bite. Fortunately, you do not have to worry about mosquitoes carrying the AIDS virus, because it is destroyed in their guts.

I will never forget the sensation of lying deathly still under my mesh net at night while conducting my postdoctoral research in the rainforests of Cameroon, West Africa, listening to the sound of a lone *Anopheles* mosquito approaching from across the room and knowing that she might carry a load of wriggling protozoans in her salivary glands. A single bite could unleash a microbial horde into my bloodstream, even in spite of regular preventive doses of liver-corroding malaria pills. One neglected gap in the fabric hung over my bed could result in a severe bout of malarial fever if I was only mildly unlucky, recurring attacks for life if I was less lucky, or painful death if this particular female happened to carry the most virulent strain of protozoan. As things worked out, after ten years of field expeditions to tropical regions around the world, I have yet to suffer an insect-borne illness.

It is hard to feel sympathetic about the lot of a female mosquito. But think about it: She is risking her life by approaching you, and she only does so for the sake of her unborn young, albeit instinctively. The risk is more than just a chance of being swatted. From her perspective, we humans are the ones who carry the parasites, and she has to gamble that you are not a carrier when she picks you out of a crowd. If you happen to be suffering from filariasis, a single blood meal from you can send a swarm of nematodes down her throat. Once inside her gut, the worms bite their way out of her stomach and burrow into the fibers of her flight muscles. This can kill her outright within an hour or two of biting you. At the very least, it must hurt, and it often inhibits her ability to fly. After a few weeks, the developing parasites burrow back out of her muscles and wriggle through her living body to her head. Sometimes infected mosquitoes sway like drunks, as the worms disrupt their brain functions; again, this may kill her outright. But in most cases the worms lodge in her proboscis and wait.

When the mosquito lands for a blood meal, the worms seem to *know*. The moment she thrusts her proboscis into your skin, they burst through the walls of her lower lip. The shock of this mass eruption often causes her to withdraw before the parasites reach your blood vessels. If she does, they may spill harmlessly onto your skin instead (lucky you). Females in this situation have been observed desperately trying to wipe their snouts clean with their feet, sometimes getting a foot stuck in the loop of a partially emerged worm.

Many mysteries about mosquitoes and black flies remain to be solved. For instance, nobody seems to know exactly how far they will fly to find you. Some species are thought to stay within a few hundred yards of their hatching place, and others are thought to travel twenty miles or more.

One thing that can be accomplished with relatively simple techniques, however, is estimating the number of biting flies in a given area. Scientists often take advantage of the black fly's ability to detect carbon dioxide in animal breath to draw them in for collection. All it takes is a block of dry ice and a suitable patch of woods. As the "ice" warms and turns to carbon dioxide gas it lures the flies in droves. I get a similar effect when I park my dark gray station wagon beside a roadside trout stream during bug season. Within moments of stopping, the black flies are there on my windshield, apparently fooled by some combination of the warmth, dark color, and exhaust gases.

My friend Lora, who works on black fly control projects in my area of the Adirondacks, once explained to me another method that her field crews use to collect data on black-fly abundance. "We use humans as bait," she said, not mincing words. "We do it eight hours a day, four times a week, all summer long. The people who do the actual measurements stand out in specific locations with long-handled nets and sweep black flies out of the air. Then we chloroform the flies, count them up, and store them in vials for future reference." For the sake of uniformity in data collection, the human bait is asked not to bathe or to change clothing during the week. She calculates black-fly densities in units of black-flies-per-minute-per-person. "On average, we get about 250 flies-per-minute-per-person."

Next time you find yourself at the mercy of biting flies, just remember Lora, or those Canadian mosquito researchers. Maybe you can then keep your wits about you long enough to conduct a courageous project of your own, keeping track of your bites-per-minute for comparison to existing standards. Perhaps you might even acknowledge a shred of sympathy for your attackers before retaliating in traditional fashion. After all, you may be facing the threat of "bug" bites, but the tiny flies are risking death for the sake of their young. And you thought that *you* had it rough?

Summer

Natives

We do a fair bit of arguing, up here in the Northern Forest. We argue about whether and how to regulate the use of private woodlands and lakeshores, whether to spray the black flies this spring, how to dispose of our garbage, and how much to salt the roads in winter. We have argued about whether we should reintroduce timber wolves into these parts or stock certain lakes with

59

trout. Some of these arguments arise from natural cantankerous-
ness and some from real conflict of interest. It is all part of living
together in a place that is neither pure wilderness nor farm nor
town alone, but a combination of many places, a landscape with
many faces and moods. Many agendas are at work here, and they
do not always mesh smoothly.

Quite a few of the current arguments dealing with land use
hinge on a central issue: who belongs here? When applied to
people, this question is usually addressed by chronicling one's
ancestry. Sometimes, the very fact that one is a "native" lends
weight to one's words, at least among other natives. But when the
question refers to nonhumans, the debate gets very murky very
quickly. Wild animals and plants do not keep written genealogies,
and they do not speak up for themselves in town meetings and
planning sessions. When the opportunity arises to try to eradi-
cate, save, utilize, or reintroduce an organism in the Northern
Forest, the resulting discussions can quickly degenerate into ver-
bal free-for-alls, rich in emotions but often rather thin on facts.

Part of the problem is that most folks do not have a clear picture of the history, the unwritten "natural history," of their region.
Older people may have memories of when times were different,
though they are not always given full credence by their listeners
(sometimes justifiably). But most of us tend to lack a solid under-
standing of where our local ecosystems have come from, in a
long-term developmental sense, and where they are heading. As a
result, we are often confused about what can legitimately be considered to be "normal" or "native" in our area and what cannot.

Let me use the issue of lake reclamation in the Adirondacks as
an example. "Reclamation" is a euphemism for poisoning undesirable fish (usually recently introduced species such as yellow
perch, *Perca flavescens*) in a lake with rotenone, then releasing a
desirable fish species (usually trout) into the lake to replace them.

There are a lot of arguments for and against reclamation, and I will not go into all of them here. But underlying this controversy is the old issue of what makes something native to a region.

In this case, the debate tends to revolve around the concept of natural (read "native") inhabitants of local lakes. Opponents of reclamation often base their arguments on the idea that we should not tamper with ecosystems and should leave the present fish populations unmolested. "Let nature take care of itself," they say. Proponents often respond that the lakes in question have been invaded by species that were not originally found there and that reclamation simply restores the original condition of a lake and protects native fish from extinction. "Let's get rid of these immigrant trash fish," they say. Both of these positions are based upon the assumption that there is a single natural condition that our lakes should ideally be in. This common assumption is fundamentally flawed, however, because lakes change from day to day, from year to year, from century to century. Who is to say which one of these countless conditions is the "natural" one?

The very idea that there is a natural state of the world (as opposed to some "unnatural" one) is overly simplistic. Change is in the very fabric of the Northern Forest and its associated habitats. Over the centuries, climates, insects, and microbes have radically altered these woods. About four thousand years ago, populations of eastern hemlock *(Tsuga canadensis)* declined rapidly in the Northern Forest, apparently decimated by insects. Since the nineteenth century, the chestnuts *(Castanea dentata)* that used to make up more than half of the forest canopy in much of the northeast vanished in a fungal plague. During my lifetime, Dutch elm disease wiped out most of the big elms *(Ulmus americana)* that used to line many roadways and village streets like majestic sentinels. And tomorrow's forests will certainly differ from today's. A fungus and insect combination is systematically eliminating beeches *(Fagus grandifolia)* from the Adirondacks, and tiny insects called "pear thrips" *(Taeniothrips inconsequens)* are threatening the sugar maples *(Acer saccharum).*

Humans, of course, have had a lot to do with many of these more recent changes. The chestnut and elm die-offs resulted from invasions of species brought to North America by people. Prior to this century, most of the biggest white pines *(Pinus strobus)* fell to the shipping industry, and many eastern hemlocks went to the tanning industry to yield up the tannins in their craggy bark. Much of the recent sugar maple decline in the Northern Forest is thought to be a response to soil degradation by atmospheric acid deposition; the resultant nutritional weakening of trees has been shown to further encourage the depredations of maple pests such as maple borers *(Glycobius speciosus)*, *Armillaria clavescens* fungi and, presumably, the thrips as well.

But let us put this into historical perspective. Twenty thousand years ago there was no Northern Forest at all, as the whole northern half of the continent sagged under a mile or more of glacial ice. Now, that is what I call a major environmental catastrophe! Dig down beneath today's forest duff, and you will find the coarse glacial sands and gravels still in place.

After the Great Thaw, roughly ten thousand years ago, lichens and mosses coated the bare, ice-scoured rock. Then shrubs moved in, followed by spruce and fir trees. Glacial lakes that once swelled with meltwater and filled glacial depressions gradually changed in size and chemistry (or dried out completely) as climates and landscapes evolved over the centuries. As time progressed, deciduous hardwood species migrated up from the south and took their places among the conifers. According to pollen studies of lake sediments in the northern Adirondacks, today's versions of forest tree assemblages date back only about three thousand years. In addition to these climatic and successional processes, fires have consumed large tracts over the centuries, further subdividing the patchwork of habitats.

You can see this characteristic patchiness for yourself in today's Northern Forest. Climb a mountain on a clear day, or take a plane ride over the woods some time when the leaves have fallen but the ground is snow-free. You can look down and identify conifer stands from above by their dark greenery and recognize hard-wood stands by their barren branches. You can then guess what the soil is like down there by seeing what kinds of trees grow on it. The hardwoods tend to cover low hills where the glaciers dumped "till" deposits that, though rocky, are rich in the rela-tively fine-grained soils that hardwoods prefer. Conifers often occupy low areas of sandy outwash left by glacial meltwater riv-ers and cling to the thin soils of rocky mountaintops. The old scars of forest fires appear as stands of fast-growing, shade-intolerant white birch *(Betula papyrifera)* or aspen *(Populus tremuloides);* these are fairly easy to recognize from a distance by the distinctive whites and golds of their trunks.

Once you get used to spotting such lingering signs of past events, you tend to see the Northern Forest as a dynamic system, as ever-changing as the rippling surface of a lake. With such a historical outline in the back of your mind, you are likely to have a hard time answering the question "What is the natural state of the Northern Forest?"

As the woodlands have changed, so have their inhabitants. The mammal communities that live here now are different from those of the past. Again, we can attribute many of these changes to our forebears. Before the hunting excesses and land clearance of the late 1800's, for example, wolves, caribou, and cougars used to inhabit these woods. Now there are no wolves or caribou, and only sporadic, unsubstantiated sightings of cougars. But if you go farther back in time, even the mammal communities of the early nineteenth century were different from earlier communities. One geologist I know found the bones of a musk ox *(Ovibos moschatus)* in old sediment deposits east of Lake Placid, New York. Back about ten thousand years ago, animals that now live only in the arctic wandered the landscape with others (such as wooly mammoths), that have vanished from this planet completely. And before their time there was only ice, for thousands of years. So, what is the natural mammal community of the Northern Forest?

Very few Adirondack lakes, though they may appear at first glance to be pristine and remote, support the same life forms that they did a hundred years ago. Many of them have been reclaimed at least once, and still more have been stocked with game fish for much of this century. You could reasonably argue that most Adirondack lakes have become glorified fish tanks. Yellow perch, northern pike *(Esox lucius)*, landlocked salmon *(Salmo salar)*, largemouth bass *(Micropterus salmoides)*, and rainbow trout *(Onchorhyncus mykiss)* were brought in from elsewhere over the last hundred years of sporting. Today, at least two-thirds of all Adirondack lakes contain non-native fish, and many in the remaining third contain fish species that are native to the region but that were introduced into lakes where they did not occur naturally. Ironically, the predecessor of the state agency that currently reclaims Adirondack lakes to remove "undesirable" perch and bass used to introduce those same species into some of these very lakes in order to establish "warm-water" fisheries. The only game fishes that the earliest human visitors could have found in central Adirondack waters were brook trout *(Salvelinus fontinalis)*, lake trout *(Salvelinus namaycush)*, and bullheads *(Ameiurus nebulosus)*. It is anybody's guess how long they took to get here after the last glaciation and what species they replaced. The supposedly natural condition of the early 1800s, to which reclaimers seek to restore these lakes, was only one of many dynamic ecological conditions.

As I suggested earlier, this all boils down to an underlying assumption about what makes someone or something "native" to a region. It is usually based upon priority of appearance in an area, but this does not generally hold up under careful scrutiny. Take, for example, the humans. Some folks who were born and raised here are still not considered to be full-fledged "Adirondackers" because their parents or grandparents came from some other place. But then consider the Mohawks, many of whom live in the Saint Lawrence River valley just north of here. They might say that any Caucasian, African American, or Asian American living here is a nonnative.

Some might say that even the Mohawks are immigrants, because their ancestors pushed earlier peoples out of their territories, and still earlier ancestors crossed over from Asia in the days of the Bering land bridge. Those Asian ancestors probably had their roots in the early hominids of Africa. If we take the priority rule to its logical conclusion, we must conclude that anybody who does not live in East Africa is an immigrant. By this reasoning, New Yorkers would be true natives if they lived in Kenya but not if they lived in New York.

The priority rule can get you into still more trouble if you take it to extremes. If you extend this principle beyond the limits of our species and continue our genealogical lineages back through the ancestral mammals to the ancestral reptiles, amphibians, fish, and various invertebrates, you may end up rooting every family tree in the first ocean-dwelling cells to appear on Earth nearly four billion years ago. This would make all nonmarine organisms non-natives. Certainly, even the native strains of brook trout swimming in our Adirondack waters arose from ancestors who lived elsewhere; they did not just spring spontaneously out of the glacial mud ten thousand years ago.

This priority standard for judging nativeness loses much of its appeal when you think about it enough. Because yellow perch have lived in Adirondack lakes for a century or so and are unlikely to vanish in the foreseeable future, perhaps they should be considered natives by now. Surely, a local lineage as long as that of the Adirondack yellow perch would qualify most humans as being natives of a region. The same can be said for certain plants, such as the lovely but generally despised dandelion, or for the starlings and honeybees that followed European humans to North America but have lived here so long that most of us now take them for granted.

The real issues in these kinds of arguments are easily obscured by the phantom issue of nativeness. I suspect that when we argue over what kinds of fish "should" be in Northern Forest lakes, we are often really disagreeing about what *want* to be there. Some of us enjoy catching trout more than perch and worry about losing unique genetic strains to extinction. Others of us prefer perch, or have no particular preference and do not want governments to spend all that money on stocking programs. Still others fear that reclamation or stocking may have unforeseen harmful effects on aquatic food chains. These are all valid concerns worthy of serious discussion, and in debating them we only get sidetracked by trying to prove or disprove the nativeness of various species. If we are really going to use the priority rule in determining lake management policy, then we might as well stock all of our lakes with primordial bacteria.

Much the same can be said for the contentious issues of wildlife introductions, timber management, and land use planning. Yes, cougars used to live here, but do we want them near our kids and pets? Yes, clearcuts are unsightly, but are they really fundamentally different from natural wildfires and blowdowns in the overall scheme of things? Yes, so-and-so's ancestors grew up here, but many outsiders have also helped long-time residents to protect this landscape from abuse, and we can all contribute to sound planning for the future.

I believe that these are the kinds of issues that we must address in order for us to coexist in this increasingly crowded Northern Forest and that it will take much careful and heartfelt thought and openness to work them out in a civilized manner. I hope that we can keep our attention focused on them and spend less time on arguments over who or what got here first.

Snakes

What is it about snakes that makes them so scary to so many people? I ask this of my Biology 141 students each year and get pretty much the same list every time: They can be poisonous. They swallow food whole. They move without legs. They are cold-blooded. They have forked tongues. As the list grows, it becomes apparent that nearly everything about snakes makes people uneasy.

Personally, I have always liked snakes, so it is a little difficult for me to sympathize. My earliest serpent encounter was with a docile ring-necked snake *(Diadophis punctatis)*, about the size and thickness of a kid's bracelet. My mother caught it in our yard and handed it to me, thus permanently defusing the Old Fear in my young mind.

But the dread of reptiles is worldwide and multicultural, permeating ancient and modern mythology and symbolism. This, and the very fact that anything that whispers *"snake"* to the human unconscious triggers a fight-or-flight response, seems to suggest an evolutionary, gut-level instinct that reduces a person's risk of being bitten by a poisonous reptile. Unfortunately for the snakes, our fear is not always focused enough to center only on poisonous types. I have met people who slaughter perfectly harmless garter snakes on sight. It does no good to point out that you are much more likely to die from a bee sting, statistically speaking, than from a poisonous snake bite in North America. Nor does it matter that the poison is mainly used to immobilize fast-moving prey for the armless, legless predator, who must then somehow get the food into its mouth and down its gullet without the help of hands or chewing teeth. This is largely an irrational fear, often powerful enough to override a head full of facts and figures.

The principal dangerous snakes in northeastern North America are copperheads *(Agkistrodon mokasen)* and timber rattlers *(Crotalus horridus)*, and neither of these is particularly common. One of my friends in North Carolina was bitten when he stepped on a baby copperhead while mowing his lawn, barefooted. His ankle swelled up, turned horrible colors, and hurt a lot for a few weeks. But he did not kill the snake that bit him, and I have always admired him for that.

Of course, even true serpent-lovers must admit that there really is some reasonable foundation to the fear of snakes. I learned recently that New York State is considering reintroducing timber rattlers into the Adirondack region. Suddenly, I had to imagine what it would really be like to have rattlers in the woods. In my woods. That would mean having to wonder what is in the wood pile, or on top of that ledge just ahead, or in the yard when the dog goes out. As with most things, it is easier to be tolerant of poisonous snakes when you do not have to deal with them face to face.

Nonpoisonous snakes, on the other hand, are well worth getting to know firsthand. They are far more common than the poisonous kinds, and spending time with one can go a long way toward easing the Old Fear.

If you live in eastern North America, the snake that you are most likely to find out sunning itself on a warm summer day is the nonvenomous common garter *(Thamnophis sirtalis)*. Garters live almost everywhere and are easy to identify. I probably should not tell you that they share the same family (Colubridae) with cobras, because garter snakes are harmless to humans. There are about a dozen garter snake species on this continent, but most are black or brown with a yellow or orange stripe running the length of the back and two more running along the sides. However, even within a single litter, the color markings can be pretty variable. One summer, in Maine, I caught a common garter that was solid black with bright red flecks along its flanks. Others report individuals with encircling bands instead of stripes. Recent studies suggest that these color patterns help the snake to evade visual predators like hawks and crows. Lengthwise striping makes it difficult for a predator to get a fix on any single body part when the snake is surging forward to escape. On the other hand, the ringlike bands found on some other kinds of snakes confuse predators most effectively when the snake moves in fits and starts.

Another diagnostic feature of garters is a red tongue with a black, forked tip. Of course, you are not likely to get a good look at those features unless you pick the snake up. Garter snakes do bite, though their teeth are not very big and they only do it when you step on them or try to pick them up. You probably will not bleed, and they almost always stop if you pretend that you do not mind the bite. The thing about garters that you do want to watch out for is a much more pungent defense against harrassment: they pee all over you. Well, it is not exactly pee, but it is yellow, gloppy, very stinky, and generally revolting. Lots of snakes, even poisonous rattlers, release foul-smelling liquids from paired glands in their vents when handled. Presumably, this is meant to discourage attackers, but relatively little research has been performed on the subject to see how well it actually works in the wild. What this adds up to is the suggestion that if you pick up a garter snake, you need not worry as much about the front end as about the rear. Make sure the tail end is well away from your pant legs until it finishes squirting.

There is a positive aspect to dealing with the vent of a garter snake, if you do not mind the risk of smelling like a reptile. It can help you tell if your captive is a male or a female. You do not have to be shy about peering too closely at the animal's privates (the embarrassing parts of snakes are internal); all you need to know is the vent's location relative to the tip of the tail. Male garter snakes have much longer tails (that is, the part between the very end of the animal and its vent) than females do, though the females tend to be longer overall. Some females can be as much as four feet long.

As long as I am telling you how much fun it is to hold garter snakes, I ought to be responsible and add a few disclaimers. I was shocked recently to discover a report in the *Journal of Morphology* entitled "The mechanism of venom secretion from Duvernoy's gland of the snake *Thamnophis sirtalis*." It seems that garter snakes, being related to cobras, actually do have rudimentary venom glands in their cheeks where their upper and lower jaws meet. When a garter bites down on something, the resultant bulging of its contracting jaw muscles can squeeze the contents of the glands out and onto its tiny back teeth. This is in contrast to poisonous vipers, who have muscles attached directly to their venom glands, the better to eject toxins. The authors were not sure whether the secretions help to subdue a victim with venom, digest prey with enzymes, or disinfect the snake's mouth, or even whether the glands are vestigial and without function.

With this in mind, you should be forewarned that there is at least one reported medical case of severe (probably allergic) reaction to a garter snake bite, so do not say I never warned you. Furthermore, it is impolite to ruin an animal's day by briefly kidnapping it. It is quite selfish and I am not trying to encourage you to do it, even though it is lots of fun, highly educational, and probably does not do any real harm to the snake unless you mishandle it.

Okay, if you insist on being so callous, why not do it right? Should you ever grab a garter snake, do not really grab it. If you clutch tightly, like a hawk or raccoon would, you are likely to terrify the poor thing or even injure it. You will certainly have a struggle on your hands (and snake squirt all over your legs), although this would probably also enable you to test my claim that garter snake bites rarely bleed. The point is not to immobilize the snake but rather to confine it delicately to your hands. Hold that snake like you would hold a baby bird, in the palms of your hands. At first, it will dart forward to escape, but all you have to do then is put a hand under its neck (you know, the part between its vent and its head) as the snake moves away from you, and keep doing this with alternating hands. The result is a sort of treadmill effect. If you do this right, chances are very good that the snake will soon figure out that you are not going to stuff it into your mouth, and it will relax. Minutes later, you will likely be able to wear it around your neck. There is no way of knowing for sure, but I could swear that some garter snakes have actually enjoyed sitting on the warm skin of my arm while I stroked them gently with a finger.

Now, about that forked tongue. Holding a garter snake is a good way to get a close look at the tongue that, to some, symbolizes all things deceitful. Some folks are afraid of snake tongues because they think they are stiff and sharp like real forks. Let one brush your skin some time and you will feel how soft it is.

It is not immediately obvious why snake tongues are forked. Aristotle wrote that the split tongue was designed to give "two-fold pleasure" from tastes. A seventeenth-century writer claimed that the tongue was for picking dirt out of the snake's nose, a sensible accessory for a groveling animal. Yet another idea was that it was split for catching flies between the tines. It was only in the twentieth century, by watching real snakes, that humans got some more realistic ideas of what snake tongues are used for. According to the latest studies, the main function of that fork is to detect odor molecules. It is a bit like pushing your nasal tissues out on your tongue to study the air instead of drawing samples in through your nostrils. By waving its moist tongue around, a snake can sweep up scent molecules for chemical analysis.

Before I go further with this chemical explanation of snake tongues, I would like to point out that a pair of researchers from Florida have recently suggested that snakes may also use their tongues to detect electrostatic charges in their surroundings. Apparently, many natural objects—including plants, animals, and the earth itself—produce positive or negative electrostatic charges. By hooking a rattlesnake rattle (minus the snake) to a voltmeter and vibrating it as a snake would, the researchers found that the shaking generates positive electrical charges of up to one hundred volts. The idea is that by vibrating its tail, a rattler could build up electrical energy in its body. Then, when the animal extends its delicate tongue towards an object, the tongue should be repelled if the object is positively charged and attracted if it is negative. In theory, moisture in the breath of hidden prey could also pick up charges from the ground and carry them downwind, along with scent molecules, to the electrosensitive tongue of a hunting snake. The proponents of this hypothesis acknowledge that it will be very difficult to test their idea in the field, and their ideas have been met with extreme skepticism by some of their peers. Still, it is an interesting idea to contemplate. And though their preliminary work has been done only on rattlesnakes, the electrostatic mechanism may apply to other snakes as well. The simple act of sliding along the ground generates enough friction to produce charges, and many nonrattlers vibrate their tails when nervous about their surroundings.

Back to the use of tongues as chemical sensors. When the snake in your hands pulls its tongue back into its mouth, it slides the odor-coated tips across a pair of pads on the floor of its mouth. When the mouth closes, these pads press up against matching holes in the roof of its mouth (Jacobson's organ) where the odor molecules on both fork tips are studied and their concentrations compared.

By waving its forked tongue around where a potential prey animal has passed, the snake can tell where the edges of the scent trail are. Garter snakes probably use this method to help them locate one of their favorite foods: earthworms. Researchers have figured out that garter snakes can use their tongues to sweep up and "taste" the minute flakes of earthworm skin that scrape off as the annelid slides against soil particles, and the snakes can follow this chemical trail to dinner. The secret of the forked tongue lies in sampling two spots at once, which helps the hunter to stay on course. If one tine of the fork finds a much lower scent concentration than the other does, the snake knows that it has strayed to one side of the trail. However, snakes generally seem to have great difficulty in telling which way a chemical trail leads, because the upstream increase in scent concentration is usually too faint for tongue-flicks to discern. Presumably, the stalker may end up heading the wrong way as often as not.

Recent studies at the University of Texas provide a clue as to how a male garter snake may get around this problem when he comes across the scent trail of a mature female during the mating season. It has to do with the way snakes move. When a snake works its way through a thicket of brush or weeds, it tends to push itself forward by pressing a body loop or two backwards against the surrounding plant stems. Because snake skin (like that of any animal) is rich in odor molecules, this leaves distinctive scent marks on the surfaces against which the skin was pressed. All that a pursuer has to do is to flick his tongue against the sides of the postlike obstacles along the trail; the sides with the strongest scent on them face the lady with that irresistible air about her.

This brings us to the matter of sex. As I mentioned earlier, snake naughty bits are well hidden most of the time, but the males do extrude their sex organs while mating. Rather than dwell on the mechanics of limbless coupling, I shall instead move on to something even more titillating: snake orgies.

Yes, some of the traditional sexual connotations that humans project on snakes are actually based in fact, at least in the case of garter snakes. When the northern spring arrives, garters emerge from their dens, where they have spent the winter entwined in a mass of snake-flesh that is believed to help reduce body water loss during hibernation by reducing the overall ratio of body surface area to volume. The males generally appear first and lounge about in the warm sunshine, waiting for action. No sooner does a mature female show up, trailing tantalizing whiffs of sex pheromones, than she becomes more than just the center of attention. She becomes the center of a writhing mass of as many as one hundred lustful males; specialists call this energetic aggregation a *mating ball*. Eventually, one of the suitors wins his love's affection, and the couple gets on with the actual sex act. This typically involves surprisingly tender foreplay in which the male presses his chin against the aromatic back of the female (that is where most of the volatile sex pheromones are concentrated) and slides his head forward until the couple's naughty bits are properly aligned. I will leave what happens in the following fifteen to ninety minutes to your imagination.

However, there is a fascinating twist to this story that is worth pointing out. In 1983, a team of Canadian zoologists intruded upon the inner secrets of two hundred garter snake mating balls and found, to their surprise, that thirty-three of the orgies involved no female. Instead, at the center of each of those balls there was something the researchers chose to call "she-males," male garters whose skin produced the same oil-based sexual attractant chemical *(vitellogenin)* that female skin does. Extract of this serpentine love potion turns male garter snakes into raving sex maniacs. Scientists have used it to persuade gullible garters to court paper towels soaked in the stuff. The she-males produce this potion and lack the usual masculine chemical, *squaline,* that would normally say "hands off" to other males (if this term had any significance for snakes).

Now, this is not to say that the she-males are into kinky behavior for the fun of it. There is no real sexual activity going on in those false mating balls. Nor are she-males attracted to males, per se. When put into a cage with receptive females, she-males act like normal males, even when males are provided as potential alternatives. In fact, she-males actually have more of the male hormone, testosterone, in their blood than normal males do. So, what are they up to? The currently accepted idea is that the she-males are actually crafty Don Juans, who boost their chances of "getting the girl" by confusing their competitors. A she-male in the mix supposedly turns he-male attention away from the real target long enough for the faker to maneuver into prime position and go for the gold.

Besides hankering after snakes that reek of femininity, male garters also seem to follow evolutionary guidelines when seeking Ms. Right. Large females are generally preferred over smaller ones, presumably because larger females produce more offspring, thus providing a maximal genetic legacy to the next generation and promoting an overall selective tendency for female garters to be larger than males. And once the act of love is consummated, the male unconsciously protects his investment in the future. The fluids in his semen coagulate in his partner's cloaca and block the entry of competitor sperm should another suitor try to follow suit.

Snakes are pretty darned interesting. If you happen to be one who hates them, I hope that these pages have helped you to appreciate their general harmlessness enough to let them go about their tongue-flicking and orgiastic escapades in peace, especially the ubiquitous garters. It is tough enough living in the wild as a limbless, fangless garter snake; the last thing they need is trouble from us.

Plant Defenses

Many of my friends are into using herb teas and herbal preparations as remedies for mind and body ailments. Some use them in moderation, more out of curiosity or a sense of fun than for anything else. Others use them as alternatives to modern medicine, preferring the essences of whole plants to "chemicals" synthesized in laboratories. This latter group tends to assume that remedies prepared from whole plants are much safer to use than are modern synthetic drugs.

I can see their point to some extent; much human suffering has resulted from the side effects of today's powerful synthetics, and they do not always cure what ails you. But the concept of plants being chemical-free and benign is rapidly becoming outdated, as scientists probe deeper into the inner workings of cellular and molecular life. Like it or not, plants have always been full of chemicals, many of them designed to manipulate, injure, or kill animals, microbes, fungi, and even other plants. After all, they cannot run away from danger, so they have little choice but to fight dirty. As a result, there are far more physiologically active chemicals present in an organic garden or on your spice rack than in the average medicine cabinet.

Once you start delving into the chemical nature of plants, you will never think of them as passive, insensible things again. As one researcher at the State University of New York at Buffalo has put it, "People just haven't really viewed plants as the dynamic, aggressive, pugilistic little beasts that they are. There's a war going on out there." With this in mind, let me take you on a tour of some of the plants growing near you, to show you what I mean. We will be exploring things that are too small to see, detectable only by the most sensitive chemical tests and gadgetry, and therefore beyond most people's awareness. I think that you will be surprised by some of the things we discover.

Let's start with quaking aspens *(Populus tremuloides)*, those lovely slender trees with the trembling leaves. Walk over to one, reach up, and pull down a leaf. Put it to your lips, take one tiny nip out of it, chew it a moment, and then spit it out without swallowing. Pretty nasty, bitter stuff. The bitterness comes from organic chemicals in the leaf, and they are there to keep you and other animals from damaging that precious photosynthetic structure. Recent investigations into the chemistry of aspen leaves reveal an amazingly intricate and effective defense system at work in that greenery. As soon as you bite the leaf, enzymes in the leaf tissues begin turning two benign substances *(salicortin* and *tremulacin)* into very toxic *phenolic glycosides.* As the concentration of toxins builds up, more of the benign substances migrate from the host twig to the leaf to keep the supply of poisons growing. Within twenty-four hours, the leaf is quite different from what it was when you first nipped it—and five times as poisonous.

How does an aspen leaf "know" that it has been bitten? There are several possible mechanisms that have been found in other plants. One way is for oxygen from the air to enter the wound and to set off a chemical reaction cascade that results in the observed toxic response. Another way is for the plant to build activating enzymes in fragile, dormant forms that break apart under the pressure of biting jaws and leap into action. Whatever the precise trigger mechanism is, the toxins that form in damaged aspen leaves are ideal for deterring animals that seek to digest them with the help of stomach acids. As soon as they land in an acidic stomach, these molecules change into an even wider array of nasties, including *phenol* and *catechol*. This should be a warning to those of you who consume an herbal remedy for the sake of a single known medicinal component; the substances in that herb could very well turn into other things once they are inside you.

This defensive capability of quaking aspen leaves is not unique. The buds use a slightly different assortment of astringent, irritating, and toxic compounds (including *benzoic* and *phenylpropenoic* acids, assorted *esters,* and *flavanones*), and different species of *Populus* use still other chemicals in their buds. One of the targets of these chemical guardians is birds. *Coniferyl benzoate* in aspen buds is very effective at repelling birds, because it stimulates their trigeminal nerves and makes them gag.

You might think that such a battery of deterrents would keep herbivores at bay. Not necessarily. Certain insects, such as larval swallowtail butterflies, take advantage of the time lag required to get the aspen defenses up to speed. They feed on a single leaf for a few hours, then move on to another one before the glycosides build up too much. And ruffed grouse *(Bonasa umbellus)* somehow manage to gobble aspen buds like popcorn, without apparent discomfort. This last observation is especially interesting in that coniferyl benzoate itself has been shown to repulse grouse, as it does other birds. Researchers in this area have suggested that different stands of aspens make different amounts of chemical deterrent and that the birds may simply be seeking out and dining on the least toxic trees. You might do well to keep this natural variability of wild plant chemistry in mind if you should ever consider dosing yourself with an herbal preparation.

These are only a few examples from a vast array of plant defenses, only a small proportion of which have yet been fully characterized by scientists. They are the legacy of an evolutionary struggle that has been going on since plant life began. As plants develop defenses, herbivores develop ways to get around them until new defenses arise. The degree of complexity in the coevolution of plants and their grazers can be quite impressive.

The bitterness that you experienced in the aspen leaf was largely owing to members of a large group of chemicals collectively known as *tannins*. Tannins not only taste bad; they also bind up enzymes and other vital proteins in your body and in the vegetation you swallow so that you cannot digest them properly, thus greatly reducing the nutritional value of your meal. The very name "tannin" refers to the ability to tan animal hides by reacting with the proteins in skin. The best-known commercial source of tannins is hemlock *(Tsuga canadensis)* tree bark, but it is very widespread in the plant kingdom, and it appears in botanical food products ranging from green tea to chocolate. Oak leaves are full of tannins, and they boost their concentrations when gypsy moth caterpillars *(Lymantria dispar)* attack them. It doesn't always work, but the caterpillars seem to enjoy their meals less under such conditions. When offered a choice between undamaged oak leaves and tannin-enriched nibbled leaves in lab tests, the bristly larvae tend to choose the former.

Gypsy moth caterpillars, nonetheless, are quite capable of de-
foliating oaks, so they must have some elegant chemical tricks of
their own to deactivate tannins. They can detoxify plant glycosides with the help of enzymes in their guts, as long as the concentrations of toxins in the leaves are not too high. Tannin-fighting
mechanisms have been discovered in mammals such as moose,
beaver, and black bears, which get around the problem with the
help of tannin-binding proteins in their saliva. However the moths
do it, they not only thwart the oak defenses but even turn them
to their advantage. Recent studies show that gypsy moth caterpillars use oak tannins in their own bodies to defend themselves
against viruses!

The next time you take a walk through a fragrant pine forest,
stop for a moment and take a deep breath. That lovely piney
aroma is caused by the *terpenes* (as in turpentine) that permeate
pine resin. Resin is not sap; it is a sticky glue that seals wounds
and gums up invading insects. The terpenes in the resin evaporate easily and flood the tunnels of wood-boring insects with
poison gas. They also deter rodents from eating the tender young
seedlings.

Plants also benefit from the effects of their own phytotoxic
("plant-poisoning") chemicals on plant competitors. Bog-dwelling
sheep laurel *(Kalmia angustifolia)* releases a host of water-soluble
compounds into the surrounding soil that damage the roots of
neighboring black spruce *(Picea mariana)* seedlings that might otherwise overgrow or crowd it. The roots of certain turf grasses do
much the same, thus tending to keep trees out of the area. And
leachate from the fallen bark, wood, and leaves of larch, spruce,
oak, and other trees helps to deter competitors from sprouting too
close to them.

I notice that my perceptions of organisms change when I spend
much time thinking about the molecules twitching within their

body tissues. After a while, I begin to think of an organism as a place, a location, a habitat. Countless molecules, each kind with characteristic structures and behaviors, occupy different habitats within the organism, some operating in relative isolation and others working in precisely tuned concert with other molecules. Never again will I think of a plant solely as a convenient source of a single spice or medicine; the diversity of molecules in any organism is staggering.

Wild plants are not the only ones that use chemical defenses, of course. Even the seemingly innocuous pollen grains of corn are loaded with poisons. If the wind-blown grains should happen to land on noncorn plants, their function in serving the home plant changes from that of procreation to that of chemical warfare. Stray corn pollen grains ooze *phenylacetic acid* and other compounds that "burn" competitors' leaves and injure the foreign flower structures on which they land, thus slowing the competition.

Potatoes are starch storage organs derived from the underground stems of plants that belong to the same family as black nightshade *(Solanum nigrum)*. Potato plants put a lot of work into producing that food supply, and they do not want any nosy animals robbing the cache. They discourage some of this theft by lacing their tubers with several varieties of alkaloid poison. One of those poisons is *solanine,* a molecule that depresses the central nervous systems of animals such as humans. Solanine is naturally present in potatoes, but its concentration varies from plant to plant. To reduce the risk of solanine poisoning, health regulations in the United States prohibit the sale of potato varieties that contain more than twenty milligrams of it per hundred grams of spud. In other words, your mashed potatoes may legally contain a measurable portion of organic poison.

The potato plant, of course, is not really being aggressive. It merely displays the unconscious process of natural selection, the

main driving force behind evolution. Those animals that eat potato (or nightshade) solanine, or the toxic *glycoalkaloids* that appear in a tuber when it turns green and begins to sprout, may be
less likely to survive and breed as successfully as those that do
not. Over generations, this can add up to a decline in the former
group and a proportional increase in the latter, to the benefit of
future potato plants.

Do you like your food spicy hot? Maybe you sprinkle red pepper on your meals. The red pepper in shakers and hot sauces
comes from the powdered fruits of the pepper plant (*Capsicum*
species). Apparently, the pepper plant's wild South American
ancestors were better off if their fruits were not eaten by mammals. Birds, on the other hand, could disperse the fruit-encased
seeds over great distances in their droppings. Producing a chemical to which mammals are more sensitive than are birds keeps
mammals from eating the fruits without scaring off the birds.

That chemical is *capsaicin.* It is one of a half-dozen organic alkaloids present in pepper flesh, and it resembles the natural hormones in a mammalian nervous system. When you take a bite of
capsaicin-laced food, the stuff binds to nerves in the tissues of
your mouth and throat and makes them send pain signals to your
brain. It also induces sweating. As in many chemically induced
nerve responses, one's reaction diminishes with exposure; this
explains why regular pepper users can down Mexican dishes that
give others nosebleeds.

How about a little herb tea to wash down your dinner? Comfrey
(*Symphytum officinale*) is very popular among herbalists and lay
folk who experiment with herbal teas and remedies. Euell Gibbons
called it "an ideal herb for making home remedies by amateur
herbalists." It has been used over the ages in Europe to treat wounds,
coughs, diarrhea, and ulcers, among other things. But comfrey does
not want you to cut it down, hang it up to dry, and crush its leaves

into tea. If you do that, comfrey will punish you with liver-poisoning alkaloids. There are several well-documented cases of liver damage caused by the ingestion of comfrey. You may not feel it right away if you use only small doses, but the damage accumulates every time you use it. By the time you notice that something is amiss, you are just as likely to blame it on the hot dogs that you ate as a child. The very fact that comfrey poultices are said to fight microbial infections in wounds should tell you that there is some property in those leaves that kills living things. People concerned about their health probably ought to steer clear of comfrey, especially if they are pregnant or nursing.

I could go on and on. There are *morphine*-like sedatives in lettuce, nasty *phenylpropanoids* in cloves, and *nicotine*, gram for gram one of the deadliest organic substances known, in tobacco. But it is not my intent to suggest that all plant chemicals are bad. In fact, plants often use chemicals that bring great pleasure and sustenance to animals—when it is in the plants' best interest to do so.

Flowers are classic examples of plants rewarding animals for services rendered. Flowers offer bright colors and pleasant odors

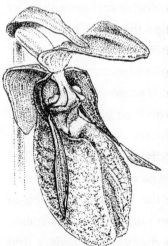

as attractants for animals that will spread pollen to other blossoms. The fragrances of flowers are gaseous chemicals of very specific kinds and combinations that serve as lures to those animals capable of detecting them. With the proper equipment, you (and wild pollinators) can tell closely-related flower species apart solely by the chemical compositions of their aromas. A team of Swedish researchers recently analyzed the fragrances of three kinds

Lady's slipper orchid

of lady's slipper orchid (*Cypripedium*

species) and found that each attracts a different species of bee with the help of either *fatty acid* derivatives, *isoprenoids,* or *phenyl* derivatives.

After the attractive aromas have done their job, nectar is the sweet reward that encourages future visits to flowers, and many pollinators live on nothing else. Chemical analyses of nectar reveal fascinating adaptations that encourage this process. The sugars in bird-pollinated flowers, for example, are often different from those in bat-pollinated flowers because of differences in the nutritional needs of birds and of bats. Vitamins and other nutritional substances often spike the sweet liquid to ensure good health in the animal associates. And, in some cases, noxious alkaloids repel generalist pollinators that may carry precious pollen off to the wrong kind of flower, without discouraging specialist species that happen to be immune to the effects of those particular alkaloids.

Fruits are another example of beneficial manipulation of animals by plants. All that sweet moist fruit flesh is not there to fertilize the ground around the seeds when it falls to earth; everything the seeds need in that regard is already present within the confines of their seed coats. The fruit is colorful and tasty in order to encourage animals to eat it. This helps the plant because the seeds are likely to pass undigested through the animal's body and to land in some distant spot in a pile of nutritious manure.

I will use a common garden tomato (*Lycopersicon* species) to show how fruiting plants get animals to do their bidding. I should point out first that tomatoes are considered to be fruits in the biological sense, although grocers classify them otherwise. The basic rule of thumb is that if it has seeds, it is a fruit. If it is a seedless domestic, just remember that the seeds used to be in there when its ancestors grew wild. Even more technically speaking, a tomato is a "berry," because it has many small seeds, a thin skin, and no core in the middle.

Now imagine an unripe tomato on the vine in your garden. How do you know it is unripe? It is hard, green, and bitter. Even nonhuman animals can tell it is not time to eat it yet. The tomato plant is keeping its animal associates at bay until the seeds are ready for dispersal. *Chlorophyll* provides the green color, *calcium pectate* keeps the cells in the fruit tightly bonded in a hard ball, and acids give the kick. As ripening progresses, *pectinase* enzymes eat away at the bonds linking the cells, turning the fruit into a soft mass. Other enzymes chop up starches into component sugars that sweeten the juices and mask much of the acid tang. Over a hundred volatile organic chemicals appear in the brew, giving the fruit that unmistakable bouquet of vine-ripened tomato.

It is now only a matter of time before something hungry arrives. If things go according to plan, it will be a roving animal, fully capable of dispersing the seeds. But it may also be a drifting fungal spore that sprouts into a tangle of threads that consume the flesh and leave the seeds to fall too close to the parent plant. How to increase the chances of attracting desirable customers? Advertise. The green chlorophyll breaks down and is replaced by orange *carotenoids*, then finally by bright red *lycopene*, which is clearly visible to color-sensitive eyes from a considerable distance.

The same basic process occurs with any animal-dispersed fruits, both wild and domestic. You could make a pretty good argument that such plants manipulate wild animals by bribing them with food for their dispersal services.

A fascinating example of this is the use of ants by wildflowers for the dispersal of seeds. The seeds of nearly one-third of all spring wildflower species in the Northern Forest are dispersed by ants, including the seeds of round-lobed hepatica (*Hepatica americanum*), wild ginger (*Asarum canadense*), violets (*Viola* species), and at least three species of *Trillium*. So dependant are these flowers on ants for seed dispersal that the abundance of ants in a given plot of woodland plays a major role in determining how many seeds get planted each year. In the case of *Trillium*, each seed is equipped with a fat-rich structure *(elaiosome)* that ants find attractive. The ants mistake the *oleic acid* in the elaiosome for the body fats in a dead animal, and they drag the whole assembly off to their underground nests to add to their food stores. Once the booty arrives at the nest, the ants feed the fatty elaiosome to their larvae and dump the seed in a refuse pile where it is fairly safe from predation and surrounded by nutrient-rich waste, an excellent situation in which to sprout.

You could also argue that our cultivars manipulate us as well. Even though we eat potato tubers and red peppers without excreting the host's seeds in a suitable manner, we do disperse and cultivate new generations to replace those that we consume. We even prepare the ground for them, water and fertilize their soil for them, and fight off their pests and competitors, often with chemical defenses of our own design. Now who is taking advantage of whom?

It adds an interesting twist to plant-animal relations when plant defenses are used for purposes other than those for which they were originally produced. The nicotine in tobacco *(Nicotiana tabacum)* leaves mimics the hormone *(acetyl choline)* that fires off nerves controlling animal muscle fibers, thus paralyzing or convulsing small herbivores foolish enough to graze on them. And yet humans consume tobacco products in great quantities, nicotine and all. Although it paralyzes the ciliated cells that clean your lungs and constricts your blood vessels (thus raising your blood pressure and reducing circulation in your extremities), nicotine rarely kills a human outright as it would a tiny insect. The dose is not usually large enough to do you in right away, although I am told that there is enough nicotine in a cigar to kill you if you eat the whole thing.

In large part, our use of tobacco arises from the addictive nature of the drug within it. Nicotine acts on animal nervous systems very effectively. In our case, it not only tells our muscles to
twitch, it also tells our brains to crave more tobacco. Thus, instead
of driving us away, nicotine drives us to consume the plant. Fortunately for the plant, we are intelligent enough to realize that we
have to sow tobacco seeds in order to replace the herbs we have
gathered (but are often too weak-willed to overcome our health-
corroding addiction). Thus, in a classic Darwinian sense, we have
become the agents of selection for the continued production of
addictive drugs in *Nicotiana tabacum*.

The chemical aspects of plant life are energizing a scientific
field that once had a reputation (among outsiders, of course) for
being rather stagnant and dull. Plants are every bit as active and
exciting as animals are, only on different scales of size and time.
As a result of this recent burst of research, it is becoming ever
more clear how complex the molecular systems making up herbs
can be and how those systems affect us. In the light of these
findings, we may need to re-examine our perceptions of accept-
able health risks as we realize that the natural world has always
been swarming with biologically produced toxins and that species like ours have nonetheless managed to deal with them quite
well through the ages without becoming extinct as a result.

On the other hand, these revelations should also serve as a
warning against careless use of chemical remedies of any kind,
even if they come from supposedly benign "natural" sources such
as herbs. Whether they are synthesized in a laboratory or in the
body of a living plant, many organic compounds can have subtle
and unexpected effects on your body. Much work remains to be
done on determining the kinds and concentrations of molecules in
even the most common species, not to mention all of the effects
that they can have on a human body when ingested by themselves

(in purified form), in concert with their numerous chemical compatriots (in whole plant form), or in fresh versus dried or otherwise prepared states. If you do use herbal remedies, even those published in traditionally reputable texts, I hope that you do so in moderation. Your liver will probably be grateful.

Fireflies

Most of us know something about fireflies and about the displays that they put on above open meadows and along forest edges on summer evenings. Some of us even get to see them up close when we take kids out to catch fireflies in a jar to make a living lantern. Fireflies give insects a good name; even people who cannot stand the sight of something with more than four legs on its body can get sentimental about them. But the lives behind the lights are interesting as well.

Fireflies are called "fireflies" because they fly around letting out flashes of yellow or greenish light. The light comes from organic chemicals—including one aptly named *luciferin* (after the Latin words *lux* and *ferre*, "light" and "to carry")—that mix in their abdomens. When the chemicals mix, they glow but do not produce much heat like true fire would. It is much like those plastic glow-sticks that glow in the dark when you whack them.

Fireflies are not the only organisms that do this. Glow-worms do it, for example, but that is not surprising because they are actually flightless female fireflies or glowing larvae. The field guides say that firefly larvae have spindle-shaped bodies and overwinter underground in swampy areas, but that little else is known about them. It is not even clear why they glow at all, because they are still too young and innocent to be thinking about things like sex. Further, some insects, including the glow-producing "fire beetles" *(Pyrophorus)*, even lay luminescent eggs.

There are many organisms other than insects that produce cold light. Certain bacteria and some fungi ("foxfire") glow in the dark. Sometimes plankton organisms, ranging in size from microscopic flagellated algae to globular, fist-sized comb jellies, can make whole areas of the sea light up at night. Free-swimming marine microbes, called dinoflagellates, are thought to emit sudden flashes of blue-green light as a means of startling predators long enough for them to escape and of attracting larger animals that might eat their attackers. Many marine fish carry phosphorescent markings on their bodies. One type *(Photoblepharon)* carries patches of light-producing bacteria living in the tissues under its eyes; it can cover or expose the patches at will, so a school of these fish looks like a twinkling city at night.

But fireflies can actually turn their lights on and off, something that many other glow-producers cannot do. On the other hand, not all fireflies produce light. Most of the species in the western

United States give up the ability to glow when they mature into adults. In the Adirondacks, I often come across fireflies crawling about on tree trunks in the full spring sunlight. They overwinter in crevices in tree bark, where many of them manage to avoid the prying eyes and beaks of nuthatches and chickadees. This species is active only during the day, presumably because the nights are still too cold for them, and as a result they do not bother flashing.

One of the surprising things about fireflies is that they are not really flies, in spite of the name. For that matter, they are not bugs, either, even though many people call them "lightning bugs." Unlike most folks, biologists are very specific about the kinds of insects to which they give seemingly straightforward names such as "flies" and "bugs." You have to look at an insect pretty closely to tell what to call it, and sometimes you have to ignore what most people call it altogether. Technically speaking, a firefly is a beetle (order Coleoptera), because it has a pair of heavy wing-covers, a pair of light transparent flight-wings underneath, and chewing mouthparts. The same goes for lady bugs; they are actually beetles, too, for the same reasons. In order to qualify as a true "bug," an insect has to have much the same kinds of wings that a beetle does, but it also has to have a sort of drinking straw setup for a mouth. True bugs (order Hemiptera) usually carry that long snout folded back under their chin, and they eat by jabbing it into a plant or animal and sucking up the body juices. Kissing bugs, giant water bugs, bed bugs, water striders, and the like are true bugs. On the other hand, to see why fireflies are not flies, take a close look at a housefly some time. True flies (order Diptera) have only the two flight-wings, with no wing covers.

Another thing most people probably do not realize is that there are lots of different kinds of firefly. You can recognize different species by looking at them up close. Most fireflies are black or brown with yellow or orange markings, and they generally range

between about ten and twenty millimeters in length. In front of
their wings is a shieldlike structure called a *pronotum*, which may
look like a broad, colorful head from above. In fact, the real head
is tiny and generally hidden under the pronotum. When the animal is sitting still, you can see its head-rooted antennae sticking
out from beneath the shield. But you can most easily identify
different species of firefly from a distance by watching the flashes
that they make at night. Stand out in a meadow some evening
when the fireflies are out and watch for patterns in the flashes.
They are not just random bursts of light; each species has an exact
pattern that it uses over and over. The patterns look like morse
code, and you can actually draw them out on paper. One might
look like a string of three or four dots, and another one might
have a long, loopy flash followed by a series of short dots. Try to
track a single individual as it flies across the meadow, and you
will see what I mean. It makes sense that fireflies are picky about
their flash patterns, because they use them to call mates, just as
birds do with their voices.

I went to a wonderful lecture by a firefly specialist when I was
in graduate school. The afternoon seminars at Duke were usually
pretty dull, to my taste. The speaker was typically some famous
person of whom I had never heard, who droned on and on about
things that I did not understand or particularly care about until
it was all I could do to stay awake. This fellow was different. In
the middle of the presentation he turned off the room lights, pulled
out a penlight, and showed the suddenly attentive audience how
to call fireflies in. All you have to do, he told us, is to learn the
flash pattern for a given species, then copy it exactly with your
penlight. If you do it just right, you can get fireflies to land on
your hand and crawl up to the penlight.

There are some fine points that you have to take into account
if this is going to work for you in the field. First, you need to

know a bit about firefly courtship. The fireflies that you are likely to see in the air are males; females usually stay on the ground. Specific habits vary with species, so I will use *Photinus collustrans* as an example. These fireflies have two peaks of mating activity, in early May and in August, when the males move into the darkest airspaces that they can find and start flashing. When the abundance of airborne males peaks, unmated females emerge from their underground burrows and watch the light show from down below in the vegetation. If a male catches a female's eye, she flashes back at him. If he sees her flashing precisely the right signal, he flies down and tries to mate with her. Recent studies suggest that at least in some species, females who fail to mate on the first night of the mating season emerge earlier and earlier on each successive evening, presumably to increase their chances of encountering a suitable male. They also tend to stay out of their burrows for longer and longer periods, even after the males have ceased signaling for the night. Why they would expose themselves to the dangers of predation in this way, without hope of finding a mate as compensation, remains a mystery.

If you want to communicate with one of those flying males, you need to copy a female's flash. In order to succeed at this, you have to be able to do the imitation with great precision. There can be several kinds of firefly using the same meadow or forest, so they are all very careful about producing and responding to the right signals; otherwise, they might land and waste valuable time and energy trying to seduce the wrong kind of firefly. Of course, it also means that they are probably going to be mighty disappointed when they land on your penlight.

The males have to juggle several agendas at once in all of this activity. Not only are they trying to impress females, they are also competing directly with each other. For instance, male *Photinus macdermotti* release pairs of flashes at regular intervals of about two seconds; interested females respond by answering about 1.2 seconds after each pair of male signals. But rival males are also watching each other, and they coordinate their displays according to what their neighboring romancers are doing. Say a male lets off a paired burst of light, and a lady responds within a little more than one second. A rival may fly in and time his flash pairs to overlap with those of the first male. While male number one is letting off his second flash, the rival lets off his first. In this manner, the interloper can sometimes confuse and distract the female enough so that she may readjust her responses to follow his flash pattern instead.

There is another good reason to be very precise in your production of and response to firefly flashes. Fireflies are predators, and they like to eat other insects. To be sure, my copy of the *Audubon Society Field Guide to North American Insects and Spiders* claims that adult fireflies do not feed at all; I assume that this blanket statement refers primarily to *Photinus pyralis,* a species whose adults, in fact, do not eat. (The larvae, though, are said to eat slugs, snails, and other insect larvae.) According to most reputable authorities, however, certain kinds of females, especially the bigger ones, like to eat other fireflies. Guess how they do their hunting. It is reminiscent of the old days on the North Carolina coast, where robbers used to go out on moonless nights and lure ships onto the deadly shoals with lamps that looked like lighthouse beacons. No wonder the males are careful. If the signal is not just right, it might be a lure to a nasty end.

There is much more going on in those sparkling fields than just a bunch of pretty little light displays. Out there among the flickering lights and chirping crickets, you can find everything from rivalry and seduction to deception and murder. The poor suitors may even fall prey to the very ladies they are trying to seduce; the females of some species continue flashing after they have mated in order to attract a postcoital snack.

Of course, just on a purely visual level, a dark meadow full of fireflies can be truly striking, especially when you see it for the first time. One of my Dutch friends just could not get enough of the sight of those flying sparks while visiting me in Maine one summer. "We don't have anything like this back home," he often exclaimed, shaking his head. "They'll never believe me."

There is still one kind of firefly display that I would especially love to see. It is one that experts had thought occurred only in Southeast Asia until a lay person with good observation skills and a healthy dose of curiosity set them straight. In 1991, researcher Jonathan Copeland got a tip from a housewife in the Great Smoky Mountains of Tennessee who had just read an article about those Asian fireflies that sit in trees and flash on and off at the same time, like bulbs in a Christmas display. She was convinced that the same thing was going on in her own woods and wondered if someone would come and check it out. When the skeptical Copeland arrived at her mountain cabin, he got quite a surprise. Copeland described the experience vividly in an interview published in *Discover* in 1994: "All of a sudden, through the trees I could see a group of fireflies going flash-flash-flash-flash-flash, and then they'd stop for a few seconds, and then flash-flash-flash-flash-flash. I thought, 'Oh, my God, it *is* there.' " The fireflies (*Photinus carolinus*) were flashing within three hundredths of a second of each other. They seemed to achieve this synchrony in much the same way that a crowd manages to clap in unison. One individual starts out with a pattern, others join in, and everybody adjusts their timing to match the group average. Copeland tested this idea by flashing lights in front of several of these fireflies so they would think that they were among neighbors. Sure enough, they adjusted their flash intervals to match his.

At this writing, nobody has yet figured out why a tree full of male fireflies would flash together like that, because the supposed point of the light display is to attract a mate. One would think that such synchrony would just make individual males vanish into the crowd. Copeland suggests that the group display may act like an extrabright beacon to call females in to the general vicinity. Once they arrive, things may change; nearby males might stop flashing and move in under cover of darkness.

Even more mysterious is the rippling wave effect that sometimes occurs when the flashing is in full swing. Neighboring groups ignite in series, one after another, sending cascades of light down the dark Smoky Mountain hillsides. Whether this is accidental or whether it is a firefly equivalent of The Wave, nobody knows. Nobody, that is, except the fireflies.

Beavers

It was late on a summer evening. I was coming home from a concert in Plattsburgh, drowsy after the long evening and the drive. I turned from Route 3 and onto the narrow country road that runs through Onchiota and marks the home stretch to Paul Smiths. The Onchiota road is flat and sparsely settled, and sometimes it invites high speeds. This was one of those times. About three miles down the dark, tree-lined road I slowed slightly to

99

navigate an upcoming turn when the car lurched to the right and suddenly decelerated. Sheets of spray roared up from all four tires, and I spun partway around in a glistening sheet of water. Fortunately, nobody was coming the other way at that hour, and I pulled out of the spin safely. It took me a few minutes to catch my breath and to realize what had happened: I had nearly been done in by the state mammal of New York.

A colony of beavers *(Castor canadensis)* had built a dam in a nearby alder swamp, and the resultant pond had flooded part of the road. Shortly after my mishap, the road washed out completely as the choked stream scoured out a new channel. Roads are scarce in this part of the Adirondacks, and such a washout meant long detours for local residents until the culvert could be cleared of sticks and the road repaired. After a few run-ins of this sort, beavers fall quickly from the status of "interesting" to that of the unprintable.

Beaver populations have expanded tremendously in the Northern Forest during the last few decades; as of 1994, there were thought to be at least eighty thousand beavers in New York State (perhaps eighteen thousand in the Adirondack Park alone). As their numbers have grown, their reputation among their human neighbors has tended to tarnish a bit. Folks who might otherwise enjoy beavers grow indignant when a prize tree is gnawed down, or real estate is flooded, or when dams drown valuable timber and inundate roads. In rural areas, few colonies get into trouble with humans, but in the more settled areas, virtually all of them do. In 1993, the New York State Department of Environmental Conservation (DEC) received more than six hundred beaver-related complaints in Saint Lawrence County alone, where colony densities average about one per square mile.

In such a context, it is hard to imagine that beavers were virtually eliminated from this state by 1900. The rise, fall, and recovery of beaver in the north country is intimately entwined with human history, and it makes a fascinating story.

The underlying theme in this story is, oddly enough, fashion. European settlers discovered that beaver fur made first-rate garments for the trendy and well-to-do. Hold a prime beaver pelt in your hands and you will know why. Long, reddish guard hairs streak an otherwise dark, dense undercoat of silky fur that can be used to make soft, warm garments. The underfur can also be combed from the hide, wetted, and processed into a kind of felt that was once used for making expensive stovepipe hats. Beginning in the seventeenth century, Native Americans were widely employed by French, British, and Dutch traders to trap and to skin the animals, and "fur wars" erupted sporadically between competing parties. So large and lucrative was the early trade in beaver pelts that a beaver graced the official seal of the Dutch colony of New Netherland on the Hudson. Skins were used as currency in New York State during the late 1600s; people called them "brown backs." Albany, today's state capital, was once known as "Beverwick."

Even so, humans apparently had little impact on the beaver population until the adoption of steel traps in the early eighteenth century, a time when logging and settling also destroyed beaver habitats all over the Northern Forest. Under this onslaught, the animals began to decline. By the mid-1800s, fewer than one thousand remained in New York.

What seemed like a miraculous reprieve occurred during the late 1800s. A semiaquatic South American rodent, the nutria *(Myacastor)*, was found to be a good substitute for beaver in the fur business, and silk came into hat fashion. By 1870, the beaver hat market collapsed. Nonetheless, beaver populations continued to decline. By 1895, fewer than one dozen colonies remained in New York, and beaver hunting was banned around the turn of the century.

Between 1901 and 1907, however, several dozen beavers were brought in from Yellowstone and other parts west and released in the Adirondacks in a desperate attempt to replenish the stocks. The timing of these introductions could hardly have been better. Enormous forest fires in 1903 and 1908 opened up vast areas of forest to colonization by aspen and birch, two of the beavers' favorite food trees. With abundant food, few large predators (wolves and cougars had long since been killed off), and little or no trapping, beavers rebounded to the fifteen thousand mark by 1915.

Besides their history, there is much about beavers that is interesting, of course, even to their detractors. Beavers are the world's second-largest rodents—after South America's capybara *(Hydrochaeris hydrochaeris)*—and can weigh over one hundred pounds, although forty to sixty pounds is the norm. Adults generally grow to about four feet in length. Their ancestors were even bigger: six-foot-long *Castoroides ohioensis* was common in New York State until shortly after the last Ice Age (roughly ten thousand years ago), and Native American legends in Maine and in the Canadian Maritimes may recall those monsters in their accounts of how the Creator shrank the original king-sized beaver prototypes down to their present form.

On land, beavers can look ungainly with their short forelegs and their barrel bellies slung low. But in water they are sleek, streamlined, and well adapted for prolonged submergence. When a beaver's nose touches water in preparation for a dive, its heart rate slows to reduce oxygen usage, and flaps inside the animal's ears seal them off from high pressures at depth. The eyelids remain open, but an inner, transparent lid *(nictitating membrane)* slides down to shield the eyeballs underwater. Extra-large lungs inflate, nostrils clamp tight, a flap at the back of the throat closes, and a thick tongue presses hard against the roof of the mouth.

Thus secure against water seepage, a beaver travels and works freely in a liquid world for up to fifteen minutes between breaths. During winter, silvery air bubbles trapped under the ice serve as wayside refilling stations. A beaver can even gnaw or carry sticks in its mouth while submerged because furry flaps of skin close behind the long, curved incisor teeth. On land, this adaptation also helps to keep the animal from choking on wood chips and sawdust while gnawing.

The flat, scaly, foot-long tail is a thunderous water-smacker that warns pondmates of danger. In hot weather, the tail is a heat-exchanger, and in winter it stores body fat. A beaver's tail is both a rudder and a driving paddle under the surface, and on land it serves as a prop while the animal sits upright for chewing.

Beavers put a great deal of care into the maintenance of their fur coats: they use their forepaws to collect water-repellent oil from a pair of glands near their vents and to spread it over the fur; split claws on two toes on each of their hind feet serve as combs to groom that soft brown pelt. Under water, those same webbed hind feet kick alternatingly or in unison to propel the animal forward at speeds of up to five miles per hour. Thus wonderfully equipped for an aquatic existence, beavers can work year around beyond the reach of most predators. This is why they build dams in the first place: to provide shelter and safe under-water access to food stores.

Beavers live between ten and twenty years and can breed by their second year. Litters of two to four kits are born in May, when the yearlings are generally driven off; if they refuse to leave, they may even be killed by their parents. However, many beaver colonies include one or more nonbreeding adults that are pre-sumably left over from an earlier batch of young.

If the dispersed young manage to avoid being eaten by bob-cats, fishers, coyotes, or other large predators, and if they can find a suitable spot, they set up shop in much the same manner that their parents did. A dam may be started with a fallen tree or be built from the ground up with sticks and mud. A pond soon forms behind the dam, and sediment and organic debris begin to accumulate on the pond bottom. Very old or abandoned ponds may eventually fill in completely and form grassy meadows in places that would otherwise have been closed forest.

The main dwelling, or lodge, resembles an igloo built of sticks, mud, stones, and leaves heaped into a mound. Lodges usually stand several feet above the water line, though the exact size and shape vary; sometimes a lodge is nothing but a hole in a river bank. The living space inside may be about three feet or so in diameter, with several tunnels connecting it to the surrounding water and with small gaps in the roof for ventilation. Sleeping platforms positioned a few inches above water level are lined with rot-resistant wood splits and shavings. Drawn by attractive accommodations, muskrats occasionally abide unmolested in beaver lodges and nibble on leftovers. But not all guests are welcome. Otters, usually satisfied with chasing down fish for dinner, sometimes enter a lodge, drive off the adults, and kill the beaver kits.

Of course, the animals must come equipped for such construction projects. Look a beaver in the face and you will see the primary tools of its trade: long, curving, orange incisors, two above and two below. If you have an opportunity to examine a beaver skull, run your fingertip lightly along the sharp cutting edges of the incisors (the word is related to "incision"). The front, yellow part is made of hard enamel, and the back part is of softer dentine. As the animal gnaws, the dentine erodes faster than does the enamel, forming a nicely beveled enamel blade. The cutting incisors grow constantly while the beaver lives, so daily chewing is a must. It is a chicken-and-egg situation; the gnawing keeps them from growing too long and piercing the animal's palate, and the teeth grow to replace what gnawing has eroded.

In the wild, keeping those teeth filed down is rarely a problem. Some experts estimate that a typical daily intake is three pounds of woody material per day (mainly bark), along with up to one pound of herbaceous vegetation. Aspen, poplar, and birch are generally the foods of choice, but beavers have been known to feed on the bark of just about any tree you can name, including resinous conifers and transplanted eucalyptus. Branches and small logs cut from the surrounding shoreline are dragged into the water for safer munching and are stacked in submerged food beds in front of the lodge for winter storage. Even beavers seem to tire of a bark diet after a long winter, though, and they often switch to greens when spring brings the ferns, grasses, and water weeds to life. Summer fare includes water lilies, berries, and even mats of algae.

As a beaver works on felling a tree, it stands up on its hind legs and leans against the trunk. It bites several times into the trunk above and below the main cut, yanks out a chip, lets it fall, and repeats. Most of the actual chiseling is done by the lower incisors, and the upper incisors provide leverage. With this approach, a six-inch birch can be felled within about ten minutes.

Beyond this, you never can tell what is going to happen to any given tree once the beaver starts in on it. Sometimes the project is abandoned in midcut because something frightened the cutter off the job. Sometimes cuts made on opposite sides of the trunk are not properly aligned and do not meet in the middle, thus failing to drop the tree. A tree may fall but get hung up on other trees, or may even land on the beaver. With a direct hit, death is instantaneous, but more often only a foot or tail is pinned, and the end is both unpleasant and long in coming.

Beavers have to be especially careful of predators when working on land. To reduce the chances of being caught out in the open, beavers often dig plunge holes connected by long tunnels or trenches to the pond. As a further hedge against predation, they also tend to do most of their cutting at night, so if you want to watch beavers in action your best strategy is probably to seek them out around sundown.

Like humans, beavers modify their habitat to suit their needs and have multiple impacts on other species as a result. As favored food trees are cut over time, shoreline vegetation shifts accordingly; in the Northern Forest, unsavory spruce and balsam fir tend to dominate shores under long-term beaver usage. The wetlands created behind dams are home to fish, amphibians, waterfowl, mink, and otters. Dead trees drowned by impoundments provide bird nesting cavities and perches, and deer and moose graze on the marsh grasses and sedges.

If you want to see a diversity of wildlife in a relatively short period of time, you will find no better place to sit and watch than beside a beaver pond. This became clear to me one spring when I hiked up to a remote pond on the flanks of Saint Regis Mountain. After following a foot trail to a stream at the base of the mountain, I came upon a series of ponds that did not appear on the map. Beavers had been at work along that stream, and some of my angler friends later told me that they go there to catch trout in the quiet pools behind the dams. A belted kingfisher (*Ceryle alcyon*) flew off with a petulant *kek-kek-kek* as I detoured around the pond. The path rose steeply from there, and the mottled green trout lily leaves that carpeted the forest floor down below became shorter and sparser, then vanished completely. As I neared the crest of the ridge, I had to slog through old snow, traveling back in seasonal time as I rose in elevation. A chorus of quacking sounds drifted over the lip of the ridge from the high-elevation pond hidden on the other side. From the water's edge, I could see the source of the racket: it was a convention of wood frogs (*Rana sylvatica*), gathered for the annual breeding orgy a full two weeks after their fellows in the lower ponds had met and dispersed.

They dimpled the water like falling rain among the stark gray trunks of drowned conifers. The cloven tracks of whitetailed deer pocked the muddy banks, and a blue heron (*Ardea herodias*) lifted heavily into the air on the far shore, mirroring itself on the still surface of the pond. To my right, I could see the stick-studded rim of the dam that had originally formed this microcosm.

It is not unusual to find beavers so high up on Adirondack mountains these days. They generally prefer low, flat areas that yield the greatest pond area for a given height of dam, but these are boom times for beavers in the Northern Forest, and it is getting harder for youngsters to find suitable damming sites in the favorable low areas. Some have even been spotted above four thousand feet in the High Peaks region of the park.

The recent beaver boom is the result of several factors. Most of their traditional predators have been exterminated, and economics and conservation laws help to keep human predation in check as well. Beaver pelts are now worth only a fraction of what they used to be, and strong laws protect many habitats from human development, leaving beavers free to exploit them to their own ends. Although the beaver's recovery from near-extinction is, in one sense, an ecological success story, you would not know it from the grumbling among north country residents. I recently attended a conference at Upper Saranac Lake in which I heard tales of culverts plugged with cornstalks gnawed from farmers' fields and of sewage plants disabled by sticks stuffed into filtration systems. I even heard a few environmentalists speak longingly of the old days of the fur trade.

Mark Brown, a wildlife biologist with the DEC, told of how unreasonably worried some people are becoming about beavers in their back yards. When faced with a four-foot rodent in your swimming pool, it is sometimes hard to remember that beavers do not attack kids or pets, nor do they eat fish out of backyard fish ponds. Another common misconception is that beavers are the sole sources of giardiasis, an intestinal ailment caused by flagellated protozoans *(Giardia lamblia)* and widely known as "beaver fever." In fact, almost any animal, from people to birds, fish, and frogs, can carry *Giardia* protozoans, and beavers should not be blamed unduly in that regard. (If you are ever suspicious of drinking water, boil it or run it through a filter with pore sizes of no more than one half of a micron).

What can we do about this conflict between humans and beavers? Live-trapping might be a solution, if only there were some place to which the nuisance animals could be moved. Longer trapping seasons have been suggested, but with pelts in such low demand it is difficult to get enough trappers interested. Land owners can plant trees that beavers find less tasty or set aside aspen or birch stands to help draw their attention away from favorite trees. Some population control programs focus on capturing beavers, injecting them with contraceptive chemicals, and releasing them.

One of the most ingenious remedies discussed at the Upper Saranac conference came from Dr. Dietland Müller-Schwarze, a prominent researcher at the State University of New York College of Environmental Science and Forestry. He advocates using the beaver's natural territorial scent marking habits to our advantage. Beavers tend to rely on their noses more than on their eyes, at least at night when they typically come ashore to work. Müller-Schwarze and his colleagues have therefore experimented with using beaver scent to erect "psychological fences" around areas likely to be invaded.

A beaver releases odor signals with the help of castor glands near its vent (or *cloaca*, from the Latin word for "sewer"), an all-purpose opening that serves as a waste exit and reproductive opening. The waxy yellowish-brown musk from these glands, called *castoreum*, contains more than one hundred chemical compounds (including alcohols, aldehydes, and odiferous phenols), is waterproof, and holds odors for a very long time; that is why it is widely used as a base for many of the fine commercial perfumes used by humans. Beavers often use castoreum to scent mark trees after they cut them, perhaps to help their compatriots to find the food source in the dark. They also set up scent posts to delineate their turf by depositing the stuff on top of little mounds of mud they place close to the water's edge. Smearing the scent on top of the mounds helps to disperse the smell farther, warning newcomers that they are not welcome. If a stranger should dare to set up its own mud mound in a colony's established territory, the residents will demolish the offensive structure and run the scoundrel off. Early in the 1980s, Müller-Schwarze's team put out artificial beaver scent marks around suitable but unoccupied habitats in the Adirondacks. They used a mix of castoreum from both males and females to suggest that mated pairs were in residence and were prepared to defend their home. Sure enough, the marked sites were colonized much less often than were unmarked ones.

Still, the quickest and cheapest way to reduce this conflict is simply to do the creatures in when they get in our way. As distasteful as it may be to those of us who abhor killing mammals (or who would prefer to see humans eliminated instead), it is an alternative worth pondering. A revitalized beaver trade could provide steady income in chronically impoverished rural economies through careful use of a renewable native resource, at the same time solving a major environmental problem. Beaver meat is edible, castor musk is perfume in the rough, and the fur is still as soft and warm as ever.

In the event that the Adirondacks should once again become a hotbed of beaver trading, I would like to pass along some potentially useful information provided to me by Bob Inslerman, a wildlife biologist with the DEC. It is an old recipe for beaver tail soup: Put skinned beaver tail into a large pan with 3 cut carrots, 2 minced garlic cloves, $1^1/_2$ cups chopped onions, and a pinch each of oregano and parsley. Cover with water, boil, and simmer for an hour, skimming off the top often. Cut tail into small pieces and put them back into pan. Add 1 cup peas, 1 cup cut potatoes, and 1 cup chopped celery. Cook another hour, adding water if needed, and season to taste.

Bon appetit!

Autumn

Mosses and Lichens

Like most people, I
grew up believing that
moss grows best on the
north side of a tree and
that if you get lost in
the woods, you can use
the mosses to help
guide yourself out. But
after about thirty years,
I finally realized that I
have never consistently

found more moss growth on the north sides of tree trunks, and I
began to wonder how the story manages to survive so well with
so little observational evidence to support it. In recent years, I
have often used this issue as an excuse to walk around in the
woods looking at tree trunks.

There are several complicating factors in gathering your own
observations regarding the distribution of mosses on trees, perhaps chief among them being the presence of mosslike lichens on

the trunks as well. I suspect that most folks do not distinguish between the two, thus confounding their observational data. And there are certainly much better methods of finding your way through the woods, such as using a compass, watching the position of the sun, moon, or stars, or following a good topo map. Still, it is a very interesting issue. The reasoning behind this hallowed claim is apparently based upon the ecological principle that different parts of a tree trunk represent different microhabitats for small organisms, presumably having something to do with moisture.

Mosses need moist surroundings in which to grow and to procreate. In part, the moisture is needed to keep them from drying out. Mosses are very small and they tend to dry out more quickly than do larger plants. Those species that live in potentially dry areas often deal with prolonged dehydration by folding their leaves close to the stems, thus reducing water loss. Another reason why mosses need abundant moisture for at least part of the year is that they need water in which to reproduce. The males release squirming sperm cells into their surroundings when the season and weather are right. Not unlike its human counterparts, a moss sperm has to be able to swim to an egg that is buried within the leafy loins of a nearby female plant and to fuse with it. The fertilized egg grows into an asexual plant that consists of nothing more than a slender stalk with a spore capsule on top of it and that obtains its nutrition from the green plant beneath it. When mature, the capsule releases a cloud of microscopic spores into the air, thus beginning a new generation of bushy little male and female plants.

Because the northern sides of trees in the northern hemisphere are the most protected from the drying effects of direct sunlight, you might reasonably expect to find more mosses growing there than on the other sides of the trunk. I have turned this problem over to several cohorts of my biology students at Paul Smith's College, and the results of their investigations are always equivocal. Part of the problem is that it is not easy to design an experiment to test the hypothesis that mosses grow best on the north sides of trees. What does "best" mean in this case? Thicker? More extensively or rapidly? Higher up on the trunk?

Even deciding what constitutes the "north side" of a tree presents a problem. Is it the whole northern half of the trunk, or just a few degrees between northeast and northwest? This can be especially important in the Adirondacks, where the prevailing westerly winds tend to trim the branches from the crowns of tall white pines *(Pinus strobus)* by piling up snow and rime ice on them in winter and breaking them off. Because of this, the western halves of such "flag trees" might be less shaded because of the lack of branches overhead, so the western half of the north side may be different from the eastern half.

After several years of field study, my students have yet to demonstrate the existence of an overall directional preference for moss growth on tree trunks, other than a general preference for growing close to the ground. Patterns sometimes appear in individual stands of trees, but when you average them all together there is no single direction preferred. Some students, in fact, have found much more evidence in favor of directional preference in the growth of lichens on tree trunks. However, lichens are an entirely different story because, unlike mosses, they are not plants.

Lichens look quite different from mosses when you get up close to them. Their colors are usually much more subdued, mostly muted greens and grays. They are incredibly beautiful under a hand lens, and are often dusted with what appears to be pastel-colored snow. The "snow" consists of thousands of minute lichen flakes that are shed continuously from the bodies of the adults; each flake is a potential new lichen.

An individual lichen is actually several organisms in one. The main body is a fungus, and the pale greenish color typical of most lichens comes from colonies of green algae or photosynthetic bacteria that live just under the "skin" of the fungus. The fungus gathers moisture and mineral nutrition from whatever it grows on, while the algae or bacteria make nutritional sugars by photosynthesis. Working together, these organisms can make a living on anything from tree bark to bare rock. The algae and bacteria that live inside lichens can grow perfectly well on their own, and scientists suspect that there are many degrees of sharing in the lichen relationship. In certain instances, all parties involved share resources fairly equitably; in others, the fungi appear to "kidnap" and to parasitize the green photosynthetic cells. The fungal components can often live without their algae as well, but they look very different when living alone. My students and I have isolated and grown some in the lab using a dish full of growth medium gel and a scalpel. We scraped off some shreds of white fungal fiber and dropped them into the dish of food, thinking that they would grow into blanched versions of the original. Instead, we grew a collection of fluffy brown, greenish, and white hairballs. This may explain why so many lichens reproduce by flaking, a system that allows them to ship several green cells off with each fungal propagule.

Late one autumn, while scrambling among open, leafless hardwoods in a gully on the northern flank of Saint Regis Mountain, my companion pointed to the tree trunks around me and said, "Look at all those strange lichens." The upwind sides of the saplings were encrusted with what looked like pastel pink house paint, from ground level to about head height. We had fun trying to explain that discovery. The prevailing westerlies probably bring more growth-sustaining moisture to the west sides of Adirondack trees and may also deposit more airborne lichen propagules for colonization. Some of the chemical compounds in our acid rain and snow might be nutritious to tree-dwelling lichens. Or perhaps the precipitation on the west sides of tree trunks causes the bark there to break down more, giving the fungal parts of the lichens more organic matter to consume?

The mystery led me to investigate lichens in the library, where I learned many interesting things. First off, I learned that many lichenologists tend to see themselves as outside the mainstream of science. Their favorite composite organisms do not fit into neat classification categories, and as a result, lichenologists do not always get invited to the meetings that algologists, mycologists ("fungologists," if you like), bacteriologists, and botanists attend each year. The good thing about this relatively marginal position of lichenology is that many of its practitioners, lacking big research grants, tend to work with low-tech equipment accessible to lay folk. The bad thing about it is that there is still not that much known about lichens.

Lichens often grow very slowly, usually by just a few millimeters a year; typical growth rates have been inferred from the diameters of lichens living on dated gravestones. Greenish lichens living on rocks in the Canadian Arctic that measure six inches or so in diameter may be three thousand years old, and some rock-dwelling lichens alive today are thought to date back to the end of the last Ice Age. Part of the explanation for this slow growth rate is that not all of a lichen's time is spent actually growing. It can go into a sort of suspended animation when it dries out or if it gets too cold or too dark for photosynthesis. Such hardiness helps to explain how they hang on to life under some of the most demanding environmental conditions available on Earth, and why there are three times as many species of lichens living above the Arctic Circle as there are of plants.

I was amazed to learn of the vast array of habitats in which lichens manage to grow. They were probably the first macroscopic life form to cover the dry land back in the Paleozoic Era, and they are still the first to colonize bare rock scraped clean by glaciers. You can find them clinging to the peak of Mount Everest, sizzling in burning deserts, freezing in central Antarctica, plastering the shells of tortoises, dangling from the tips of conifer branches in bogs, etching the glass on cathedral windows, staining the spattered rock perches of mountain rodents, flopping in the swift currents of rivers, and carpeting the backs of unfortunate insects.

There are even lichens that penetrate the very fabric of rocks, in the form of delicate fungal fibers. The fibers slip into the microscopic crannies between mineral grains and somehow collect enough diffuse sunlight to survive. Land snails in Israel's Negev Desert grind away the surfaces of stones to get at the fungal fibers within. The only time you can spot these kinds of lichens easily is when they reproduce. To do that, they form lumpy fruiting bodies, like the ones you see in the center of any crusty tree-clinging lichen, out on the rock surface where they can scatter spores on the wind. All you can see on the rock surface is the dots of the fruiting bodies, like the grin of the Cheshire Cat.

People from every corner of the globe seem to have used lichens for something or other at some time or other. Finns extract antibiotics from them; penicillin, after all, is a chemical released by *Penicillium* fungus to wipe out bacterial competitors. The French make soap out of them. Japanese boil them into soup, Arabs bake them into bread, and Siberian monks brew them into beer. Some Egyptian mummies were packed in lichen stuffing. Greeks and Scots stain fabrics with lichen dyes, and many scientists measure acidity and alkalinity with paper soaked in litmus, one of the old lichen dyes.

But what about directional preferences in lichen growth? I never did find a definitive answer in my search of the literature, but I got close a few times. To begin with, different species of lichens colonize different parts of a tree in the vertical plane. The upper-most canopy lichens get more sunlight and direct precipitation than do those lower down, and they probably depend on it. You might think that the bottom dwellers would like to live up there, too, but some species actually shrivel and die if you transplant them to slightly higher elevations on the same tree. As far as I could tell, much work remains to be done before we fully under-stand the vertical distribution of lichens in forests.

Lichens are also choosy about the kinds of tree they colonize. Some grow only on conifers, others on certain species of oak tree. This makes sense because different trees have different growth forms, with attendant differences in exposure to the elements, and they also differ in the chemistry, texture, and water-retaining capacity of their bark.

Air pollution wreaks havoc on many kinds of lichen, so you do not often find a great diversity of them on skyscrapers and urban monuments or downwind from factories. *Flavoparmelia*, a leafy lichen common on tree trunks in Ohio that is normally pale green, turns yellow and begins to die as seasonal concentrations of Cin-cinnati air pollution rise. Lichenologists have solemnly recorded the disappearances of hundreds of lichen species from cities in Europe and in North America since the Industrial Revolution. The problem seems to be that lichens take up whatever falls on them, be it water or pollutants. Because they do not shed leaves or excrete as do animals, they appear to have no way to get rid of undesirable chemicals. Eventually, toxic buildups kill the ur-ban lichens.

If you stop and think about what it takes to kill a lichen, this air pollution stuff can be pretty scary. About one hundred years ago, a French scientist tested the mettle of *Xanthoria* lichens under unbelievably harsh conditions. He kept them in a vacuum for eight years, and they did not flinch. He stuck them in a deep freezer at minus 459.9 degrees Fahrenheit, just above absolute zero, for two weeks. No problem. With that in mind, look at the record of lichen growth around factories, such as the one in Kvarntorp, Sweden, that opened in 1942. The resident lichens vanished in short order, returning only after the factory closed in 1966. I now look at the lichen-encrusted branches of the woods around my Adirondack home and breathe a bit easier knowing that they would not be there if the air was not fairly safe for me to inhale, at least for the time being.

I am sorry that I cannot tell you exactly why mosses and lichens grow where they do, or even on which side of a tree they grow best. Next time you think you might get lost in a forest, you had better bring a compass (and remind yourself, before you go out, which end of the needle points north). Save the mosses and lichens for biologizing, unless you enjoy unexpected sleep-outs in the woods.

Bogs

Ask people what they think of when they think of bogs, and you
will probably hear a list of unpleasant sights, smells, and other
sensations. To many, bogs are places to avoid, fetid pits of slime
and muck where bugs swarm and disgusting creatures slither.
But for those who are lucky enough to know a bog personally,
such negative associations are rare. Bogs are far from slimy. If

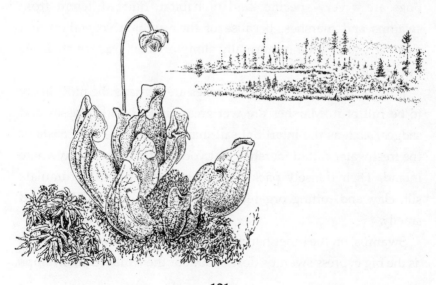

they have an odor, it is usually one of sweet, mossy growth, and not much reptilian or amphibian slithering goes on there. Granted, the bugs can be pestiferous in late spring and early summer, but they are generally no worse than they are elsewhere in the Northern Forest at that time of year.

And if you have the slightest interest in plants, bogs are a botanical paradise. Rare and delicate orchids, such as the calypso *(Calypso bulbosa)* and the white bog orchid *(Platanthera dilata)*, reward explorers of northern boglands. Shrubby Labrador tea *(Ledum groenlandicum)* sports leaves with curled edges and fuzzy white undersides that turn brown with age and that help to conserve water. Delicious ground-hugging cranberries *(Vaccinium)* thrive underfoot in the wettest wetlands. Indeed, plants are the star attractions of bog habitats, because animals are relatively rare there. But that is not to say that these are wholly placid, docile gardens. Look around a bog with a knowledgeable eye, and you will find mosses that ooze acid and plants that eat meat.

Part of the public relations problem with bogs is the result of terminology; they are often confused with other kinds of wetland. Bogs are a very specific kind of habitat, quite different from swamps and marshes. Because of their unique vegetation and chemistry, they are not normally slimy or smelly, as we shall see in a moment.

If it is slime that you seek, marshes are perhaps the most likely to be full of it. Marshes are wet areas choked with grasses and sedges, such as the intertidal salt marshes that line our coasts or the freshwater cattail marshes that you can find almost anywhere inland. Their densely packed plant stems trap and accumulate silt, clay, and rotting organic debris that can be both slimy and smelly.

Swamps, on the other hand, are mainly wooded wetlands, such as the big cypress swamps down south or the white cedar swamps in which Northern Forest deer like to shelter. You might be able to dredge some pretty rank-smelling organic slime from the bottom of a swamp.

But bogs are wetlands packed full of and maintained by semi-aquatic mosses of the genus *Sphagnum*. *Sphagnum* is easy to identify, with its dense, tattered-looking head of leaves, usually bright green or the color of red wine. Its stem is soft and floppy, so the plant falls over unless supported by neighbors, and the wispy branches covering it are rough with fine leaflets. Sometimes you find little seedlike structures attached to the growing moss heads. Those are capsules that disperse reproductive spores on the wind. But a *Sphagnum* plant can also reproduce by breaking pieces off from the main stem. As a result, the tightly packed mosses in a bog mat are often genetically identical clones.

The most effective way to fall in love with bogs is to walk on one. I do this with my biology students every year in a small wetland near campus. It is not exactly a classic bog, such as the quaking bogs that bounce underfoot like trampolines. It is thick with larch and spruce that make it seem, at a distance, to be more of a forest than a bog. But once you set foot on it, you can see that the trees shelter a thick blanket of *Sphagnum*. Quiet pools of tea-colored water glisten in low spots among the trunks and fallen logs, surrounded by soft green billows of moss.

We have marked off a single footpath to minimize impact on the place, and I enjoy watching the students' reactions when they take their first steps on that path. The first response is usually "EEEEEEUUUWWW!" as fancy running shoes press deep into wet moss. But folks soon realize that they are not going to sink up to their ears in goo and that they are surrounded by a magical world of strange, lovely, and fascinating plants.

When you walk on a *Sphagnum* bog, water squishes out from beneath your feet with a gurgling sound. Many of my students are surprised that the water looks clean rather than slimy; any debris floating in it tends to be fibrous and easily brushed off. That is a hint about the unusual chemical ecology of *Sphagnum;* it loves acidity. Dip a pH probe into the bog water soaking into your shoe and you may find a reading close to the acidity of lemon juice. Most aquatic organisms would die in that stuff, but *Sphagnum* thrives in it.

In fact, *Sphagnum* helps to make bog water acidic in the first place. The mosses soak up charged nutritional mineral ions through tiny spaces in their leaves and stems and incorporate them into their bodies during growth. But in order to get many of those ions out of solution, the moss cells have to release something in return. Hydrogen ions, so small that they consist of nothing but single, positively charged protons, leak out of the plant cells to take the places of the larger positively charged nutrients in the surrounding wetness by a process known as "ion exchange." These ions flood the waters of the bog, making it highly acidic; the term "pH," after all, is a contraction of "potentia Hydrogenii," the abundance of hydrogen ions in solution. The more hydrogen ions that occupy a given volume of water, the more acidic it is (and, confusingly, the lower its pH is).

It is a wonder that the acid buildup does not pickle the *Sphagnum.* Well, actually, it does. Reach down and pull a single plant out from among its neighbors. The top portion is lush and alive, but the straggly, long-dead bottom portion is pale and emaciated, virtually mummified by the acids permeating its peaty grave. *Sphagnum* grows ever skywards like a slow comet, trailing its past behind it.

The acidity of bog water pickles most living and once-living things quite effectively, and that is the principle behind its relative cleanliness and the fibrous texture of the underlying bog mat. The mat is mostly an accumulation of dead *Sphagnum,* much of it ancient. Under normal circumstances, worms, bacteria, and fungi would reduce such plant debris to organic mush. But these are not normal circumstances, and most agents of decay stand little chance of survival in the acid bath. Most of the decay in peat bogs occurs in the uppermost inches of the bog mat, above the level of the acidic water table.

When a *Sphagnum* plant dies, its carcass persists for years. Shove your hand down into the soft, cool thickness of a bog mat, clench a fist, and pull it back up. Chances are good that your hand will come up coated with crumbly *Sphagnum* body parts and your fist will clutch a wad of pale, dripping fibrous pulp. Put some of that pulp under a microscope, and you will see moss leaves, spores, stem sections, and shreds of well-preserved cellular tissue. By contrast, if this were a bacteria-riddled marsh or swamp, your hand might come up covered with organic glop and smelling like rotten eggs.

Give a clump of bog moss a squeeze, and water trickles out. The body tissues of *Sphagnum* are very porous. Empty cellular chambers in the leaves and stems, large enough to see clearly with a hand lens, store water for protection against dry spells. With such storage spaces, some species of *Sphagnum* can hold up to twenty-five times their weight in water. These mosses are so good at hoarding water that the surfaces of many bogs lie several feet above the normal water table, yet they remain well soaked and can even maintain open water in surface pools. Part of the reason for this is that rainfall is quickly captured and absorbed into the nether reaches of the bog mat. Another reason is that capillary action draws water into the tiny intercellular spaces in the moss leaves and stems, overcoming the force of gravity to wick water up from below. These "raised bogs" are like thick sponges lying flat in a shallow pan of liquid; their tops are moist even though they lie above the original water level. The great water-retaining ability of bogs helps them to survive sporadic droughts. When rain is scarce, effluent streams may shrivel and vanish as the bog witholds its stored water. The mosses on the upper surface might have a hard time of it, but many of them are adapted to just this sort of stress.

There are more than 50 species of *Sphagnum* in North America, many of them limited to specific bog microhabitats. Some species, like *Sphagnum recurvum* and *S. cuspidatum,* live only in the wettest spots, such as watery depressions. Others, such as the tiny, ruby red *S. rubellum* and brown *S. fuscum,* form hummocks that rise well above the surface pools. The depression-dwellers tend to lie flat and to keep most of their bodies in contact with low-lying water. Many can survive complete dehydration should their water-home evaporate, quickly reviving when wetted again. Oddly, many of the hummock-dwellers are less able to survive drying. They stand upright and clump tightly together to help them to retain water more effectively and to resist fatal desiccation.

Humans have long put *Sphagnum* to many uses. Many homes in the British Isles are still heated with dense brown bricks of peat, the compressed remains of ancient bog mats. The pungent aroma of burning peat permeates the air of Scottish towns, and it brings a distinctive fragrance to the single malt Scotch whiskeys that are brewed from grains that have been cooked over smoky peat fires. Native Americans living in bog country sometimes packed their moccasins with *Sphagnum.* Try it some time in place of socks, if you have a pair of old footwear with which to experiment. It is soft but supple and forms neatly to your foot, and it keeps your toes surprisingly warm. Just dry a mess of *Sphagnum* in the sun, then pack it in.

The acidity of bogs makes them wonderful archives of the past. Just about anything that falls into a Northern Forest bog will probably stay there until the next Ice Age. You can use pollen trapped in peat layers to reconstruct ecological histories of local vegetation and of climate change. Scientists have unearthed the pickled bodies of humans in British bogs that date back thousands of years. An entire mastodon carcass was dug out of one North American peat bog; the beast was so well preserved that its last meal of spruce boughs still filled its stomach.

Only a little less dramatically, you can get a sense of the great depth of time represented in a bog with the help of a long, pointed pole. Brace your feet, press the point of the pole into the peat mat, and see how far down it goes. If you feel strong resistance, then you may have hit something solid down there, perhaps a log that fell thousands of years ago; just move over a bit and try again. A pair of students in my biology class once pushed a metal pipe ten feet down into our campus bog without hitting bottom. Some bog mats are more than forty feet thick. Peat accumulation rates vary a great deal, but a rough rule of thumb is about one century for every one to three inches of peat.

Now, about those plants that eat meat. Carnivorous plants really make you sit up and take notice of bogs. There is something awry in the very idea of an animal-eating plant, a deliciously horrifying concept. But, in a bog, things are not normal. Plants do not normally thrive in strongly acidic waters; it hurts their roots and makes it difficult for them to extract water and nutrition from the sodden soil. Furthermore, mosses bind up most of the available nutrients in the bog for their own use and release them very slowly when they die, because of the scarcity of decomposers. Ironically, many bog plants must struggle vigorously against drought and starvation.

Under normal conditions, much of the fertility of soils originates in the bodies of dead organisms, the natural compost of the planet. In the bog, though, carnivorous plants lay claim to those nutrients bound in animal bodies before the bodies actually die.

The classic carnivorous plant, the Venus's flytrap *(Dionaea muscipula)*, is native to the Cape Fear region of North Carolina, but it is so uncommon that I never found any during the ten months I once spent in their native heartland. The meat-eaters in the bogs of the Northern Forest, the sundews *(Drosera)* and pitcher plants *(Sarracenia)*, are much more numerous.

Sundews are the smaller and the more widely distributed of the two; I first encountered them on the margins of an abandoned clay tennis court in Maine. There are five species of sundew in the northeastern United States, but you are probably most likely to find the round-leafed sundew *(D. rotundifolia)*. All five species bear the distinctive red hairs that secrete a sparkling transparent glue and that give these plants their name. The main function of the hairy leaves is to do the job that is normally assigned to roots: soaking up nutrients and distributing them to the rest of the plant. Anything that lands on a sundew's sparkling leaf, be it a hunting spider or a patrolling fly, is quickly ensnared. Chemical sedatives mixed into the goop may further reduce the chance of escape. The hairs bend slowly inward from all sides to dab their sticky tips against the target, and the entire leaf pad may enfold it like a hot dog roll. Then follows slow digestion by microbes and by enzymes in the glue until the transfer of animal molecules to plant cells is completed. Eventually, the leaf unfurls, the hairs stand erect, and the empty husk of what once was a tiny creature falls to the ground.

Sundews often respond differently to different kinds of prey. If you drop a dead mosquito onto a sundew leaf, the hairs curl around the body so slowly that the process can take days to complete. But if you lower a live mosquito softly onto the waiting glue bed so that it lays there struggling, the plant seems to know. Sundews have been reported to wrap up live prey within twenty minutes.

There is a sort of fitting irony in the sacrifice of a mosquito to a sundew. The mosquito sought your blood to stock up on protein for the production of her eggs; sundews also use their prey for reproductive nutrition. If no animal protein comes their way during the growing season, the plants can propagate by budding. But to produce the seeds that normally develop in autumn atop a slender central stalk, a sundew must capture and digest animal prey. The more insects that are available, the more flower stalks can be produced and the more nutrition is left over to support new leaf growth the following spring.

Unlike sundews, pitcher plants live only in bogs and only in North America. There are seven species of pitcher plant on this continent, but only one lives in the Northern Forest; *Sarracenia purpurea*. A pitcher plant has several large, curving, horn-shaped leaves radiating from a central hub. You may also find a fleshy, burgundy-tinted flower hung like a parasol from a central stem rooted in the hub. That is a good way to spot the plants at a distance, because the green of the young leaves tends to blend in with the background vegetation.

The edges of the leaves of a pitcher plant are folded over and fused together into a keel that runs the length of the leaf, forming an inverted cone that traps rainwater. A flaring lip at the top of the cone is coated with a fine fuzz of downward-pointing hairs. Bright red veins wrinkle across the green background, and sweet nectar oozes from glands scattered across this lip. That is the visual and gustatory bait that lures insects in for the kill. Intrigued by what appears to be a nectar-filled flower, an insect enters the trap. A short way down, the cone narrows and the walls become smooth and slick with sticky, loose cells. The victim, its feet caked with plant cells, cannot climb back out easily, and even if it has wings it has trouble navigating the constricted zone. If all goes according to plan, one more insect will soon be struggling in the pool of rainwater at the base of that green chamber of death. Plant-generated chemicals in the fluid help to break the surface tension of the water and thus to soak the insect more readily, making it harder for it to float. Aquatic bacteria soon compost the soft innards of the drowned insect, and some may generate nutritious compounds from the atmospheric nitrogen dissolved in the pitcher plant liquids. When the flowering season arrives, the plant also releases digestive enzymes into the brew. Cells in the base of the pitcher absorb the bug juice and distribute nutrients throughout the plant. Leftover indigestibles, such as chitinous exoskeletons, sink into the narrow shaft at the very bottom of the leaf until the leaf shrivels with the onset of cold weather.

In spite of the pitcher plant's reputation for dining on insects, some insects use them for food, shelter, and nurseries. *Exyria rolandiana* moths lay their eggs on pitcher plant leaves. When the caterpillars hatch in spring, each crawls off to claim a single leaf for its own. Each caterpillar lives inside the leaf pitcher, grazing on the inner lining and sometimes weaving a silken web across the opening, presumably to keep predatory wasps and other undesirables from entering. In autumn, the damaged leaf forms a protective case for the moth pupa to overwinter in. Similarly, *Wyeomia smithii* mosquitoes lay their eggs in pitcher plant pool-traps. When the wriggling larvae hatch, they somehow manage to resist being digested, perhaps with the help of chemicals that counteract the plant's enzymes, and grow to adulthood in their watery hideout. They are not always alone in there, however. An unusual kind of fly (*Sarcophaga*) spends its larval life crawling around in the edible bug debris at the very bottom of the pitcher.

If you would like to explore a pitcher plant's interior, you can easily do so without damaging the leaves. Just dip a turkey baster or other suction device into it and draw up a sample of fluid. Squirt it out onto a shallow dish and scan the treasure with a hand lense.

I hope you take the opportunity to visit a bog some time, if you have not done so already. But if you step out onto a bog, remember that your footsteps may remain a long time after you leave; please limit your wanderings to a small area. Be ready to sink a leg, without warning, into a soft watery spot in the bog mat. And remember, should your collector instincts begin to stir, that many bog plants are protected by law. They are always most beautiful and interesting in their natural surroundings; please leave them for the next person to enjoy and for the next generation of bog plants to follow.

What Should You Do When You See a Bear?

My first encounter with an Adirondack black bear in the wild took me by surprise. I was driving past Lake Clear on my way to work one morning when a large black hulk lumbered across the road in front of me and vanished into the undergrowth. From behind the wheel, I reveled in that brief glimpse of wildness.

Black bears embody much of the romance of the Northern Forest, a lasting reminder of earlier days when other large predators roamed these woods as well. Their ability to subsist on a wide variety of foods has helped them to withstand hunting pressure and human encroachment more successfully than did the wolves, wolverines, and cougars that were driven to local extinction during the nineteenth century. Even so, you do not see black bears very often. They are usually very timid around humans, and under normal circumstances you are much more likely to encounter indirect signs of bear presence than to see the actual animal.

Black bears *(Ursus americanus)* are quite different from others of their kind. When we speak of bears in the Northern Forest, we are not talking about enormous predatory grizzlies *(Ursus horribilis)* or man-munching polar bears *(Ursus maritimus)* that sling two-hundred-pound seals over their shoulders like lunch sacks. Rather, black bears are the ones that guide books call "generally harmless to humans" and "omnivorous" (they will eat your sandwich whether it contains beef or bean sprouts). An autumn-fattened female can weigh 160 pounds, and a male can tip the scales at 300 (the record male in New York weighed 750 pounds). This is far less than the weight of a western grizzly, which often exceeds half a ton. But if you unexpectedly meet a black bear in the woods, you quickly notice that it still outweighs you, and it is difficult to forget that a small but significant number of black bears—all presumably well accustomed to human presence (and garbage)— have been known to attack, kill, and even eat an occasional hiker.

One spring morning I went fishing with my friend Phil. We spent the morning in his canoe drifting down a sluggish tributary of the Saint Regis River, following a narrow winding corridor of overhanging alder thickets as we plucked tiny brook trout from the dark current with our fly rods. The stream we rode mean-dered through broad open expanses of burned land, now covered with lush tangles of bracken fern and blueberry. Sometimes when we rounded a tight curve we sent whitetailed deer crashing away through the undergrowth. We pulled out at Cranberry Rapids, where our other vehicle awaited us on a dirt track that runs through the barrens. "I like to jog out here a lot," Phil mused as we lifted the canoe onto the truck bed. "But sometimes I get nervous running through this place alone. One day I almost ran into a bear that was eating blueberries on a little rise beside the road. Usually when you see one out in the open like that, you mainly get to watch its rear end as it runs away. But this guy was different. He just stood up on his hind legs and stared at me. He wasn't going anywhere, not with all those berries left to eat." It is unlikely that a black bear in this situation would attack a human, but that fact is little comfort when you face one alone and unarmed except for your water bottle. "Basically, he got to watch *my* rear end in retreat that time."

Just beyond the barrens, Phil suddenly stopped beside a telephone pole and stepped out of the truck. The pole wore a ruff of foot-long splinters at about head height, just below a "No Trespassing" sign. "This wasn't here before," Phil said as I joined him beside the pole. "Looks like something's been sharpening its claws on it." I walked up to press my fingers into a deep gouge in the dense wood. Alongside the groove, entangled in a spray of slivers, two long black hairs swayed in the breeze. "Must have been marking his territory," Phil muttered. I wondered if it was the work of that same bear who had defended his hill so defiantly— and if that clawed message was intended for Phil.

I once attended a public slide presentation on the subject of bears by Lou Berchielli, a wildlife biologist with the New York State Department of Environmental Conservation. Lou began by pointing out that the average black bear encounter is more of a privilege than a hazard: "Bears usually see, smell, or hear you long before you even know they're around. They know that we humans are predators, and they steer clear of us. You're lucky if you get to see anything more of them than tracks."

Bear tracks are easy to recognize in the soft, fine-grained soil of a stream bank, footpath, or dirt road. Each pawprint sports a row of scratch marks dug by five toes tipped with nonretractable claws. According to Lou, it is also relatively simple to tell male from female by the sizes of the prints: "If the heel of the hind print is more than four inches wide, it's a male. If the heel print is smaller than that, it's either a female or a yearling."

Not only are wild bears typically shy of humans; they do not seem to bother other large animals very often, either. Lou offered an amusing example of this: "Black bears in these parts often seem to be more interested in insects than in larger prey. A few years back, I examined a road-killed deer that a bear had been feeding on, expecting to see large pieces of the deer missing. But that wasn't the case. All the bear had done was to claw open the carcass to get at the fly larvae and carrion beetles inside it."

In addition to their utility in feeding, bear claws—essentially just elongated toenails—make passable tree-climbing tools. You can often find old claw marks on tree trunks if you look carefully. They are easiest to spot on trees that have smooth, light-colored bark, especially beeches (*Fagus grandifolia*).

Normally, changing day length provides the instinctive trigger for denning in black bears, but in years when beechnut crops are particularly heavy ("mast years"), bears often go into their winter dens about a month later than usual to take advantage of that food bonus. Females are also thought to give birth to larger litters in the winters following heavy beechnut yields.

You can pretty much forget about seeing bears in winter, although they sometimes wake up and step outside for a brief stroll. I have seen their tracks in January powder in northern Maine, but that is rather unusual. Although a black bear's seasonal sleep is too shallow to count as true hibernation in a technical sense (Lou Berchielli calls it "periodic sleeping"), bears do experience a drop in body temperature of five to ten degrees Fahrenheit during the winter. Because their body temperature is still relatively high during this down time, it is important for bears to conserve body heat as much as possible. This is probably why their dens tend to be quite small and easy to keep warm in. A typical rock crevice or hollow tree den may measure only one or two feet in height or width. Some dens, however, consist of nothing more than a pile of brush, and some bears have even been found spending the winter curled up right out in the open.

During their long sleep, bears lower their metabolic rates along with their body temperatures, a physiological adaptation that helps them last up to five months without eating. They also gain some nutrition from their own urea by diverting it from their bladder to their intestines, where bacteria convert it into a form that a bear's body can use for building vital amino acids and proteins. And because they do not take in food all winter, hibernating black bears do not mess their beds. A fibrous plug of leaves, pine needles, and fur swallowed just before bed-time corks them quite effectively.

In the case of females, though, their sleeping bodies do occasionally emit other things: half-pound, nearly naked, newborn cubs, and the milk on which to suckle them. There are usually only two cubs in a litter, and they instinctively nestle into mom's thick fur to get at her teats and to snack until spring. A mother bear may or may not be awake at the actual moment of birth, but she often rouses enough to roll onto her back or to sit up and help her newborns to nurse.

Black bears do not normally eat for two weeks or so after emerging from their dens (late March for males, early April for females), but when the animal's digestive system eventually kicks in, the first thing it usually wants is salad. Wet areas are a prime destination then, because some of the earliest spring plants thrive there, such as skunk cabbage and marsh marigold. Later, forest glades and fields feel the press of heavy feet as bears arrive to graze on grasses, dandelions, and wild strawberries. During warmer months, you might come across a rotten stump or log that has been torn apart where *Ursus* has excavated grubs, beetles, ants, or crickets. And when the blueberries, cherries, and raspberries ripen, you can tell that bears have been at work in the berry patches by the crushed foliage and by the large droppings peppered with undigested seeds.

Bears also feed in even less elegant ways, and the surest way to see bears in the Adirondacks used to be to hang out at the town dump. But several years ago our public landfills were closed in favor of recycling and trash export, and local bear society apparently went through a period of upheaval as result. For about a year after the closings the dump bears, descended from a long lineage of trash scavengers, began showing up at our back doors.

One summer morning in 1992, a bear broke into a house just down the road from my place and emptied my friend's refrigerator while she was at work. My neighbors and I began to wonder if we might be safer living in the inner city. Over in Oswegatchie, about an hour and a half west of Paul Smiths, my fiancée's family's summer cottage was also burgled. The place was almost untouched except for one broken window, its jagged frame edges flecked with crimson droplets and black hairs, and for the remains of a single glass vase that once held a bouquet of fragrant phlox. As we figured it, the bear must have simply followed its nose to the faint scent of last year's flowers.

The rest of that summer did still less to arouse my sympathy for displaced bears. I grogged awake one morning to the sound of heavy thumping at my back door. Then I heard a crash, followed by the sustained tinkle of falling glass. Turning to look through my flimsy bedroom window screen, I beheld an enormous hairy black bear behind making its majestic way into the woods, leaving a broken porch window and a shattered sense of security in its wake. And the next day, the college dean nearly collided with a large bear while driving past the campus quadrangle. She screamed, and the bear bolted; in its terror, it crashed headlong into the chain link fence of the tennis courts, fell, got back up, and lit out for my hillside. All I heard of this commotion was the loud bang of a flying impact against the metal garbage can beside my back door. After my experience of the previous morning, I decided it was a good time to stay in bed.

But one person who lives near me actually encourages bears to visit his house. Ray Fadden is a retired schoolteacher who owns and curates a roadside museum in Onchiota and who is well known in the region as a tireless advocate of Native American rights and culture. It is less widely known that he also feeds bears at his house. This all began when a nearby dump closed and the bears were left without their familiar food source. The kind-hearted old man took it upon himself to look after them and has been at it ever since. Friends often show up on Ray's doorstep with sacks of sunflower seeds for the chickadees and chipmunks, or with trash bags full of raw animal parts. I brought seeds last time.

"I've got one youngster this year who's been causing trouble at night," Ray said as we left his back door. His unkempt shock of white hair blazed like a beacon atop his red flanneled shoulders, and I followed it along a path that winds through the woods behind his house. "He keeps breaking in through the screen door to our porch. Most of them wait outside to be fed." Ray did not seem to be worried about his own safety as he told me this; he was more concerned about the cost of fixing that screen door. I got the feeling that I was listening to a father complain that his teenage son had just batted a baseball through another plate glass window.

The heavy odor of rotting flesh sweetened the air as we approached one of the feeding stations. It was just a stump among the trees, but the ground was well packed around it, and a chorus of flies hummed in the undergrowth. They sounded happy to hear us coming. "I leave food out for the bears all summer, and they show up almost every night. But when hunting season starts,

I come out here and fire off a single rifle shot. The bears don't come around any more when they hear that. They know it's not safe in these woods then."

I have never sat out at night to watch the bears come to Ray's feeding stations. They know him better than they know me, and I am not sure how comfortable either of us would feel out there in the dark together. Besides, it strikes me as a rather personal, private connection that the old man and those bears share, not the sort of thing upon which outsiders should intrude.

Sad to say, Ray's bears have more to worry about these days than they used to. Not only are the dumps closed and sport hunters hot on their trails each fall, some folks in the Far East believe that black bear gall bladders cure impotence and are willing to pay a thousand dollars or more for a single bladder, enough to touch off a booming illegal trade. Fortunately, gall bladder poaching is less common in New York than in other states, in part because it is legal to sell those body parts here, so they yield much less profit than do the black market prices.

In spite of hunting, poaching, and car collisions, New York's bear population seems to be in little danger of extirpation. We are their only predators; there seems to be plenty of good habitat for them; and because they eat such a wide variety of food, they are not likely to starve if one or more food sources dries up now and then. And it is probably just as well for the bears that our local landfills have closed. Garbage actually comprised only a fraction of the diet of the so-called dump bears, because they generally filled up on wild foods growing near the dumps. Although most of their trash-foraging was apparently out of curiosity and a taste for exotic snacks, it was probably harmful in the long run. Bear droppings around landfills were often full of paper from diapers, paper towels, and picnic plates, and on at least one occasion, an individual was seen with indigestible plastic strips dangling from its rectum. Dump bears also drank heavily from toxic leachate puddles.

Because black bears are pretty common in the Northern Forest, it is probably a good idea to make a list of do's and don'ts for an encounter with *Ursus americanus*. Overall, the best rule of thumb is to use common sense in your dealings with them. For example, avoid doing something silly like spilling tuna oil all over yourself, like I did during a camping trip through Montana's grizzly country in 1976. Do not eat in your sleeping bag, for similar reasons. Rather than invite break-ins by keeping food and other smelly things (including toothpaste) in your tent, stash them outside in a plastic trash bag hung well out of reach on a sturdy line strung between two trees. This will help to keep marauding raccoons out of your gear as well.

As long as we are on the subject of camping in bear country, you may be glad to know that an actual scientific field study purports to have shown that black bears are not attracted to human menstrual odors. Rather than get too graphic about the methods involved in that study, I will just reassure you that nobody was staked out as bait.

Of the few bear attacks reported in the Northern Forest, the vast majority have involved a good measure of human folly. For instance, one teen had his butt bitten when he tried to make off with two cubs; a pretty reasonable and restrained response from a concerned mother, I dare say. Just remember that a bear is not going to act like a human in your presence. It is more likely to treat you like another bear. That may involve swatting you hard with a powerful clawed mitt or giving you a good chomp if it thinks you are getting out of line.

In the unlikely event that you find yourself facing an unfriendly-looking bear (ears laid back, hackles raised, jaws working), I am sorry but I am not sure what to tell you. After searching through several dozen sources, I have collected a perfectly confusing array of encounter strategies from which to choose: stand your ground, back slowly away, lie still, clap your hands, talk gently, and scream your head off while charging the beast. Trying to flee is also reasonable if you have a good head start, but black bears have been clocked running at thirty miles an hour, and they climb trees better than you do. If you are actually being chewed on by a black bear, experts generally consider it to be a good idea to holler and fight back rather than to lie still, because the animal is clearly under the impression that you are suitable food and requires some convincing to the contrary.

So, what should you do if you see a bear in the Northern Forest? For those who choose to avoid this issue altogether, just hang your food at night, keep a clean camp, and make some noise occasionally, and I can almost guarantee that you will never see *Ursus americanus* anywhere but in a zoo, a picture book, or on a roadside. For the rest, I recommend standing very still and savoring the moment as best you can. Before you know it, the animal will probably run for cover and leave you with nothing but a lingering shot of adrenaline and a good story to tell.

Conifers

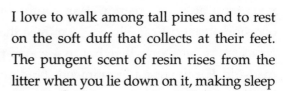

I love to walk among tall pines and to rest on the soft duff that collects at their feet. The pungent scent of resin rises from the litter when you lie down on it, making sleep come more easily. And when you awaken from your nap, the dark green crowns lean over you, their sway- ing tops almost touch- ing in the blue space above. In winter, the stunted red spruce (*Picea rubens*) and balsam fir (*Abies balsamea*) that cling to windswept mountaintops are caked with heavy crusts of rime ice and may bear horizontal icicles that point upward. In calm weather, snow also builds up on their icy boughs like creamy wax drippings. Seeing them in that condition, I am always amazed that they can survive up there.

To me, these kinds of trees personify the Northern Forest. The classic Northern Forest scene is one of dense, dark stands of conical evergreens with their ragged spires etched against a brooding sky, or perhaps reflected on the mirror-smooth surface of a cold northern lake. Even for people who do not live in the north, evergreens are heavy with symbolism; what is Christmas without a fragrant tree or evergreen wreaths?

Needle-bearing trees are among the easiest for beginning naturalists to recognize and to name. They are also among the most interesting trees in the Northern Forest, once you get to know them a bit. For those not yet familiar with them, these trees are full of surprises.

One of the first surprises we need to deal with here has to do with names. We really should not be calling them evergreens in an exclusive sense; the term "evergreen" refers not just to needle-bearing trees but to any plant that stays green all year round. This means that mosses, ferns, club mosses, many herbs, and most tropical forest trees are evergreens, as well as the Christmas trees that we usually have in mind when we use the term. We are going to have to be more specific.

Besides, needle-bearing trees are not completely evergreen, either. The needles do not last forever, and a certain percentage of them are shed each year when the rest of the forest loses its leaves. The flat, aromatic needles of balsam fir, for example, usually last about eight to ten years, so you can expect a fir to drop about one-tenth of its foliage every autumn. The needles that fall off are situated farthest from the elongating twig tips, where this year's newest growth looks a brighter green than do the older needles, so a fir branch gets proportionately more barren as it ages. Balsam fir, by the way, is the tree you are most likely to use as a Christmas tree, largely because its resin smells so good when the needles dry out. Folks also chop up its needles and stuff balsam pillows with them for year-round fragrance.

Most other trees commonly referred to as "evergreens" hold their needles for much shorter periods. Eastern hemlock *(Tsuga canadensis)* also has flat needles, but they are shorter than those of fir and they stay on the twig for only three years. White pine *(Pinus strobus)* needles are long, thin, and angular, and they cluster together in clumps of five. White pine needles last only a couple of years before browning and dropping; if you look closely at a twig just behind the oldest green needles, you can see dozens of little bumps where earlier needle clusters once grew. But the least evergreen of the evergreens is the bog-dwelling larch *(Larix laricina)*, whose sprays of short, floppy needles turn golden and drop every autumn. My students always get a chuckle when a friend or parent comes up from the city in October to visit the Adirondacks and says, "Isn't it just awful how the acid rain up here is killing all those poor evergreens across the road?"

The proper way to describe what most folks think of as evergreens is to refer to their private parts, the cones. That is how plants are usually classified, after all, because reproductive structures have to be unique to each plant type in order to produce the same species with each successive generation. So you might want to replace the term "evergreen" with "conifer," to refer to trees that reproduce with cones. As long as you stay away from the tropics (where you may encounter palmlike, cone-producing cycads), the term "conifer" will evoke the desired images of green needles, woody cones, and ragged skylines.

Conifers are truly outstanding trees, even though they are thought to be in something of an evolutionary decline in terms of species diversity. The world's tallest and oldest trees are all conifers. The tallest California redwoods *(Sequoia sempervirens)* soar more than three hundred feet above their massive, buttressed root systems. California also boasts giant sequoias *(Sequoiadendron giganteum)* that measure more than thirty feet in diameter at the base and that can be more than three thousand years old. The famous General Sherman tree is 275 feet tall and 32 feet thick at the base. Even its branches are enormous: the *Textbook of Dendrology* that I have on my desk says that you can lie crosswise (yes, crosswise) on the largest branch of the General Sherman tree and not be visible from the ground, 130 feet below. The world's oldest known trees are the twisted, blasted-looking bristlecone pines *(Pinus aristata)* of the American Southwest; growth rings counted in core samples from some of these ancient beings yield ages of up to five thousand years.

Once people get used to talking about conifers, a final obstacle to correct classification is the confusion most of us get into over the word "pine." Show somebody a flaky brown conifer cone and they are likely to call it a pine cone. It may very well be a pine cone, but it is important to remember that not all conifers are pines. Here in the Adirondacks, some of our most common conifers include balsam fir, red spruce, and eastern hemlock, as well as white pine and red pine *(Pinus resinosa).* So the thing you have just picked up may be a fir, spruce, or hemlock cone rather than a real pine cone.

Luckily, the cones of different conifers are easy to tell apart by their characteristic sizes and shapes. Hemlock cones are tiny—about the size of your thumbnail—and shaped much like pistachio nuts when their scales are still tightly closed. Fir cones are shaped like elongated footballs set upright on the parent branches and are about as long as your pinky finger. White pine cones are among the longest in these parts, measuring four or more inches in length. Many of them curve slightly, and their big, resin-tipped, platy scales gape like clamshells to release the winged seeds that develop at the base of each scale.

But even with the cones there are surprises. When we talk about white pine cones, for example, we often assume that white pines make only one kind of cone. In fact, they make two kinds, because pines are sexual and produce sperm and eggs in order to reproduce. The big cones that we see squirrels picking apart are female cones, in which microscopic "egg" cells develop into seeds after being fertilized with sperm carried in by airborne pollen grains. The pollen itself forms within a completely different kind of cone on the same trees, one that is so small and crumbly that you rarely see it on the ground. These "pollen cones" (as opposed to the female "seed cones") throw great clouds of yellow pollen dust into the air in spring or early summer. On windy days, sudden gusts can make the limbs of tall conifers smolder with the stuff. The pollen rain can get so thick in piney woodlands that the air is visibly clouded by it. Fortunately, those of us who suffer from allergies are rarely as sensitive to pine pollen as we are to that of ragweed or of grasses.

The purpose of a conifer pollen grain is to carry a pair of gene-toting sperm nuclei to a young seed cone, where it forms a root-like pollen tube that pierces the woody heart of the female cone and delivers the sperm. One sperm fertilizes an egg cell that will become the seedling; the other fertilizes a pair of cells that will proliferate and form the starchy food supply that makes up the bulk of the seed. Thus a seed is born.

Because the pollen has no way to move on its own, much less to aim for a proper target, pine pollen dispersal is a shotgun affair designed to counteract the overwhelmingly poor odds of success with overwhelming numbers of randomly drifting grains. The seed cones of many conifers also help the pollen to succeed in its mission; individual seed cone scales are often tipped with dabs of sticky resin that may help pollen grains to stick tight once they land. Conifers depend on the wind rather than on animals to carry their pollen to waiting seed cones, so each grain is outfitted with a pair of gas-filled bladders that help it to float on air currents. Because the bladders make the pollen very light, it tends to float when it first lands on water. When the pines are spewing pollen, entire lakes can turn yellow, and winds push the drifting grains ashore in thick, pasty windrows that some mistake for chemical pollutants.

I remember walking along the shore of a remote lake in Siberia, near Lake Baikal, in August 1990. The dark, rounded boulders lining the shore were zebra-striped with dozens of perfectly parallel yellow pollen bands. It was an unusually hot, dry summer, and the lake had shrunk steadily for several months. After the Siberian pines finished unloading their dusty propagules in early summer, the falling lake levels had painted a bright pollen stripe at each successive step in their decline.

That trip to Siberia opened my eyes to a global perspective on pines and other conifers. Although overall numbers of conifer species have been declining since they reached their zenith in the Mesozoic dinosaur era, the few remaining species persist in staggering abundance. There is a nearly continuous band of predominantly conifer forest, called *taiga*, that encircles the globe around forty-five degrees north latitude. If I had flown due east from Lake Baikal, I could have reached the Adirondacks without ever leaving the taiga belt (except, of course, while crossing the Bering Strait).

Sadly, this ecological link to the rest of the northern midlatitudes is being eroded as foreign nations scramble to cut as much Siberian taiga as possible while the sagging Russian economy cries out for quick cash. Perhaps even more disturbing, however, is the widespread trend of conifer decline caused by atmospheric pollutants. Controversial at first, the idea that sulfur and nitrogen oxides from the burning of fossil fuels could kill forests is now widely accepted among scientists.

The Adirondacks have the unfortunate distinction of representing one of the classic examples of conifer decline resulting from atmospheric fallout. Pollutants blow in largely from Ontario in the winter and from the Ohio Valley in summer, and they fall on these mountains as their parent air-masses rise with the topography, cool, and drop condensing moisture. Red spruce and balsam fir that cling to the thin rocky soils along the tops of Adirondack peaks and ridges are at greatest risk, and not just because of the harshness of climatic and soil conditions up there: most clouds stay above three thousand feet elevation here, so trees growing above that height spend a great deal of time engulfed in cloud. Unfortunately for the trees, clouds in today's mountains are often full of nasty pollutants, including heavy metals, PCBs, and corrosive ozone. They are also acidic. The fog condenses on the conifer needles, runs down the branches and trunks, and drips to the ground. The acidic fog water (or melting rime ice, in winter) damages the protective waxes on the surfaces of the needles, exposing them to dehydration, and changes the chemical compositions both of the foliage and of the bark. Once in the ground, the acids also wash nutrients away and mobilize aluminum and other substances that damage the roots and their associated mycorrhizal fungi, thus reducing mineral and water transport to the already stressed needles. Such unhappy plants are prime targets for frost and drought damage, not to mention fungal and insect attack, and there are places in these woods where it seems like every red spruce has browning branch tips.

We tend to associate pines and other conifers with cold climates, but I should point out that pines do very well in hot, dry climates as well. The coastal plain of the southeastern United States is thick with loblolly *(Pinus taeda)*, longleaf *(Pinus palustris)*, and other pines. Ponderosa pine *(Pinus ponderosa)* thrives on arid mountain slopes in the western states. In any climate, it is drought resistance more than cold tolerance that gives pines an edge over other plant competitors: conifer needles are coated with waxes that help them to retain moisture, and their roots are hooked up to mycorrhizal fungi that help them to extract as much water as possible from the soil. You might not normally think of drought when you see a wintery landscape. But, as far as the trees are concerned, that snow might as well be parched Sahara sand. The underlying ground water has turned to stone and the trees' sap transport vessels have frozen as solid as water pipes in a poorly insulated cottage.

Winter's cold, of course, is also a problem in and of itself for northern conifers. Although trees do not generate body heat, they do make physiological adjustments as day lengths shorten and as temperatures fall. This process is called *hardening*. Trees that have not undergone hardening are likely to die if temperatures dip below freezing for very long. James Halfpenny and Roy Ozanne, in their wonderful book on winter ecology *(Winter: An Ecological Handbook)*, describe the hardening process very well. They divide the complex transformation into distinct steps that occur just before and after the onset of cold weather.

In early autumn, when squirrels are storing nuts and bears are fattening themselves for the winter fast, the conifers prepare for winter in their own ways. Hormones pass from the needles to the stems and the trunk. These chemicals make the cells there more permeable to certain substances, halt growth throughout the tree, and shunt carbohydrate food reserves into storage zones in the roots. When this first stage is completed, a tree can survive temperatures that sink into the teens. This is largely because the permeable cells of the trunk and the stem have absorbed substances that act as antifreeze. The dilute fluids around individual cells may solidify at such temperatures, but not the cells themselves.

With the first frost, the next stage of hardening begins. Changes that occur on the molecular level throughout the living tissues further strengthen the tree's freeze resistance. Vital proteins are surrounded by protective sugar molecules that keep freezable water molecules at a safe distance. Fatty cell membranes are rebuilt to accomodate water loss caused by freeze-drying. After this stage is completed, the tree can endure subzero temperatures without much harm to its cells from ice expansion or from the dehydration that occurs when ice crystallizes in interstitial fluids and draws water out of the surrounding cells. Dehydration, in fact, helps the cells to resist freezing by leaving their contents more chemically concentrated than before.

Even though a conifer's needles are shut down for the winter, they are still green and they can still trap sunlight. This can become a problem over time, because destructive oxidizing molecules may form as a result of too much sunlight capture and may wreak havoc among the inner workings of the needles, especially when most of the rest of the plant's cellular machinery is in cold storage. Thus, part of the hardening process involves producing protective antioxidant molecules in the needles.

After all this preparation, you might think that a pine tree would be pretty well set for whatever winter could throw at it. This is generally so, but sometimes hardened trees are disastrously confused by brief midwinter thaws that can "deharden" them for a day or more, then swing suddenly back to frigid temperatures before the trees can regain their freeze resistance. It is not so much a problem of extreme cold as it is of sudden cold. Some trees, once they are properly winter hardened, can withstand a dip in liquid nitrogen (minus 196°F); but midwinter warming followed by sudden cold can be disastrous, because hardening is lost faster than it is gained. Several years ago, we had one of these dramatic temperature shifts in January. The mercury stood at minus 25 degrees when dawn came to the weather station atop Whiteface Mountain; by midday, it had shot up to 40 degrees, but by midnight the air temperature plummeted back to minus 25. The winter burn on conifers that can result from this sort of thing resembles acid rain damage; needles on the branch tips turn brown and die, in part because ice crystals have burst their cell structures.

It is a pleasure to savor the beauty and the ecological strategies of northern conifers, but you can also put them to practical use. The resin that oozes from wounds on red spruce trunks hardens into amber lumps that you can break off and chew like gum (particularly useful if you have any old fillings you want to get rid of). I have to admit, it tastes terrible at first, but if you tough out the first five minutes the turpentine flavor leaches out and you are left with a mouthful of surprisingly pleasant chew.

The soundboard on my Martin guitar is crafted from Sitka spruce *(Picea sitchensis)*, and resonant spruce wood is an important component of violins and many other stringed instruments. Years ago, Adirondack spruce was the luthier's tree of choice, until overharvesting claimed the richest stands. Now Sitka's valuable trees face a similar fate.

In the old days, spruce was as important as white birch in the construction of bark canoes, because the delicate craft were framed with spruce wood, laced with spruce roots, and sealed with spruce pitch. Native peoples in the Northern Forest also used the root lacings and pitch in bark basketry. I once fashioned my own rough version of the watertight baskets that Wabanaki people used to make, while I was camping at Moosehead Lake in Maine. I cut a sturdy sheet of white birch bark *(Betula papyrifera)* from a large windfall and pulled yard-long rootlets from the duff of a nearby spruce grove. I scraped the cortex from the roots with a jacknife, leaving a dozen strong, pliable lacings, each about as thick as a tooth on a pocket comb. After lacing up the basket, I sealed the seams with pitch prepared according to an old voyageur recipe. It worked beautifully.

To make waterproof pitch glue out in the woods, boil some water over a campfire in an old tin can and drop in a few chunks of hardened resin. A clear sticky fluid soon collects on the surface of the water. Skim that off onto a sheet of birch bark and stir in a mess of powdered charcoal from last night's fire until the mix turns thick and black. Smear it onto whatever you want glued, and let it stand overnight. The glue is strong but rather brittle, and because it is black it is clearly visible on a birch-bark background, so in these times of Elmer's and epoxy, the old stuff is probably best suited to camp emergencies and woodcraft purposes.

When I take my biology students into the woods for their first outdoor lab session, their primary assignment is to identify all of the species of conifers that grow along the trail. The students learn that white pines are the tallest of all, soaring a hundred feet or more above the forest floor. They discover how prickly spruce needles are, figure out that fir needles are longer than the hemlock needles they otherwise resemble, and try to remember that the sticky stuff on the conifer trunks is not sap (that would be watery) but resin that heals wounds and repels insect invaders.

But the crowning experience of the day comes when we eat the conifers. Well, actually, we drink them. When we reach the lean-to on the shore of Osgood Pond, all of the green conifer twigs that we gathered along the trail go into a tin can full of boiling water. After steeping for ten minutes, the brew is ready to be poured off into mugs and lightly sweetened with honey. This is where students fresh from the city have to face a dilemma, as youthful intellect meets gut instinct. Should they believe this professor when he reassures them that eastern hemlock is not the poisonous plant that killed Socrates? Most of them bravely opt for a taste, and an ancient tradition is passed on to a generation that may never again live among wild conifers. One hundred years ago, lumber-jacks in the Northern Forest drank conifer needle tea as a source of scurvy-fighting vitamin C, as well as for pleasure. And they learned about it from the Native peoples who knew these woods so well.

I never tire of watching young forestry students savor their first steaming mugs of spruce, pine, fir, and hemlock needle tea. There they stand, in the heart of a grove of wind-bitten red pines on a sandy ridge beside a glistening northern lake. They lift their cups and sip gingerly, then stand quietly to take in the rich evergreen aroma with a distant look in their eyes.

Princess Pines

Each autumn, when the trees have lost their leaves but the ground is still free of snow, I like to take a walk in the woods just to look for green things growing on the forest floor. Many plants up here in the Northern Forest stay green throughout the year, and they are not all conifers.

One November morning I followed the trail that weaves through a lovely patch of woods near my house. The treetops were barren of foliage, but the forest floor still bore vestiges of summer's greenery. The bases of the tree trunks were thick with green moss carpeting. Here and there among the trunks were ground-hugging ferns and delicate wintergreen plants with those mealy red berries that taste like candy. Most striking of all, though, were dense patches of dark green plants that resembled miniature palm trees, spread profusely among the fallen leaves. These plants go by many names, but one of the most common is "princess pine."

Most people do not notice princess pines, especially when the leaves are still on the trees, and the ground-dwellers blend in with the overall lushness. They look so much like small pine seedlings, or big bushy mosses (those types would actually be called "club mosses") that you are likely to pass them by without realizing that they are a totally different kind of plant.

There are nearly one thousand species of princess pines and club mosses around the world, but they are all closely related, fitting under the title Lycophyta. They resemble most plants in being equipped with water-transporting veins (mosses lack such plumbing). But many lycophytes also resemble mosses at first glance because their stems are densely covered with prickly little green leaflets that can make them look scaly or almost bushy.

The green branchlets and stems of princess pines stay fresh-looking all winter, and they stand out prominently on the dry browns of a forest floor. Being evergreen like that may have contributed to the name, even though princess pines do not make pine cones. People often collect the tough, pliable plants and make Christmas wreaths and lush table arrangements out of them. They last a long time that way, despite the dryness of life on bare walls and tabletops. All you have to do is soak them in water for an hour or so to revive them. Although princess pines do a fine job of evoking holiday spirit, I do not want to encourage you to go out and collect great heaps of them every year. So many people are doing it already that some of these plants are in danger of being wiped out.

This evergeen aspect of princess pines is probably a link to their distant Devonian origins, back before the time of the dinosaurs. Between about 350 and 250 million years ago, swampy areas in the tropics were full of princess pine ancestors, all of which were probably evergreen and many of which were huge. One such lycophyte (*Lepidodendron*) from that time period had a trunk that was up to three feet thick and was split into two main branches about thirty feet up. These continued to a total height of one hundred feet or more. There were smaller lycophytes, too, although you would never know it by the sizes of their names. Take *Baragwanathia*, for example (and forget the pronunciation). This was one of the first lycophytes ever, and its stem was about one inch thick. Two-inch needlelike leaves bristling out all over it made it look more robust and, for that matter, very much like an enlarged version of some modern club mosses. Many of those long-vanished lycophyte forests are now compressed into the coal seams of Illinois and parts of the Appalachians.

Because the lycophytes seem to have changed little since their origin in the tropics, it should not be surprising that they can stay green all year. Many tropical forest plants hang on to their foliage all year long and tend to have waxy, drought-resistant leaves to help them make it through the dry season. This may explain why princess pines survive so long while hanging on your door as well as how they resist drying out while winter locks the Northern Forest's ground water in an icy embrace.

Another adaptation that helps princess pines to survive harsh northern winters was discovered only recently: they use antifreeze. Specialized compounds known as "thermal hysteresis proteins"— not unlike those in certain cold-climate fish, insects, and flowering plants—build up in the moist tissues of the princess pines and resist freezing. A final chemical adaptation, an array of alkaloids, probably helps discourage grazing from animals. There is a long and growing list of such chemicals found in princess pine tissues, including *dihydrolycopodine*, which may help deter insects and other creatures from eating the leaves.

Interestingly, it is these very compounds that have led humans in cultures around the world to use princess pines and their relatives in a wide array of traditional herbal remedies. A recent issue of the *Annals of Internal Medicine*, for example, described a *Lycopodium*-based herbal remedy that has been used for centuries in

China as a treatment for insomnia and as a painkiller; it has also appeared recently on the lucrative herbal remedy market in the United States, under the name "Jin Bu Huan Anodyne Tablets." Although *Lycopodium* may well have the aforementioned effects on humans, the point of the article was to warn doctors that it also causes acute hepatitis. Although herbal remedies are widely assumed to be harmless, this newly discovered liver-toxicity in *Lycopodium* (and many other widely used herbs) should not be surprising, because plants do not produce medically active chemicals simply for the benefit of humans but rather as defenses against being eaten.

Sometimes I like to imagine traveling through time when I am out in the woods. At night, I may look up at the stars and think of how old their light is. Some of those stars may not even exist any more, but there is no way to tell, because all we can see is the light that they released millions of years ago. Looking at the night sky is a way of gazing back through time, but you can do the same thing during the day by looking at princess pines. Lie down on your belly beside a patch of them, and you will settle into a scene that has been around for hundreds of millions of years. There are flattened princess pines embedded in Illinois coal that look just like the ones growing on top of those same fossil deposits. We humans like to think of our great cultural traditions as being ancient; our habit of cultivating plants, for example, may be as much as ten thousand years old. But set that against an unbroken line of genetic tradition rooted four hundred million years in the past. That impresses the heck out of me. Princess pines developed a superb set of features during the early stages of land colonization and have apparently not had to adjust them much since then.

If you have ever pulled a princess pine up by the roots (something that I do not recommend doing), you probably yanked up a whole clump of plants at once. Princess pines are often linked by long, rootlike runners just under the surface of the ground. What looks like a line or cluster of princess pines is often a string of genetically identical shoots growing out of the same runner. Because they are all attached to the same tissues, you might think of them as a single sprawling plant with many limbs rather than as a group of individuals. Technically speaking, the plants in that clump are clones, as are certain ferns and mushrooms, quaking aspen groves, and sassafrass thickets.

Princess pines depend upon their subterranean connections to older root systems when water is scarce, as is often the case during a dry summer on sandy northern soils. Roots as old as eight years pass moisture along to green bushy tips growing more than a foot away. Most of a newly sprouted princess pine shoot's body moisture comes from older root systems. These elder roots, perhaps three or four years old, act as a sort of base camp supply depot for pioneering shoots that can use their pooled resources to spread across inhospitable bits of soil or of shade that would otherwise be barriers to growth.

Another thing that might catch your eye is the presence of what appear to be slender candelabras or cylindrical clubs growing out of *Lycopodium* tops in the late summer and fall. You do not see that sort of thing on most plants; they are dispersal towers for dusty yellow spores. The towers start out green and sausage-shaped, then turn flaky and golden when the spores are ready to scatter on the wind. Give one of them a flick of your finger in early autumn, and you will release a puff of what looks like yellow smoke. Princess pine spore dust was once used as flash powder, in the early days of photography. The tiny particles, being composed of flammable microscopic flecks of organic matter, blaze like coal dust when ignited.

Most plants reproduce with seeds, each with an embryonic plant and starchy food supply encased in a tough coating. Spores, on the other hand, are something you might normally associate with mushrooms and mold, just tiny capsules containing solitary embryos. Spores have to land on some sort of food source, like a rotten log or a heap of fertile leaf litter, if the embryo is going to sprout and to survive. That is why spores are made in huge numbers, to make up for the lousy odds of a drifting spore landing on a suitable spot. A few other plants, such as ferns and mosses, reproduce with spores as well, but true pines reproduce with pollen and seed cones, not free-floating spores.

So, princess pines are not pines at all. They belong to their own unique division of the plant kingdom, and they are about as different from true pines as dogs are from crickets. Several of the princess pine relatives you might come across in the north woods have equally misleading names, such as "shining club moss," "ground cedar," and "running pine."

Princess pines are pretty surprising little plants, all in all. They even have sex lives. It all begins with those spores. When a princess pine spore sits on the right kind of forest floor, it sprouts into an odd little plant that lives on or under the decaying litter for ten or fifteen years and never develops much in the way of leaves. Botanists call this miniature growth form a *gametophyte* because it reproduces with sperm and eggs ("gametes") rather than with spores. Even more oddly, these gametophytes come in two very different forms. In one instance, the spores sprout after a few days atop the leaf litter, and produce tiny green, fleshy, disk- or top-shaped gametophytes. In other situations, the spores delay sprouting for two or three years, gradually washing down deeper and deeper into the forest floor. When they finally get around to sprouting, fungal fibers mingle with their tiny rootlets and help them to draw sustenance from the rich organic mulch surrounding them. Because they are so deep in the dark soil, there is not enough light for photosynthesis, and these gametophytes are not green. The little plants are so dependant upon the fungi for nutrition that they die if they do not hook up with them.

Now, to tell the truth, I have never seen one of these gameto-phytes. I do not even know anybody who has seen one in the wild, though it is not for want of looking. Somebody, somewhere must have seen them, of course, because botanical texts do men-tion this Jekyll-Hyde transformation in princess pines, as well as in ferns, horsetails, and mosses.

Then again, you should not necessarily believe everything that you read in natural history books. Just a few centuries ago, sober European botanicals claimed that ferns have invisible flowers. Back in those days of unicorns, burnings at the stake, and papal indul-gences, the intellectual ancestors of today's botanists assumed that all plants have flowers. Because none of them had ever actually seen fern flowers, their reasoning went, the things must be invis-ible. Nobody had as yet paid much attention to the little brown spots on the undersides of many fern fronds, where the spores hang out. So, the way I see it, if you want to doubt the existence of princess pine gametophytes until you actually see one for your-self, go right ahead. Meanwhile, I will just assume that they are invisible.

Some lycophytes (*e.g.*, *Selaginella*) produce two sizes of spore; the smaller spores sprout into male gametophytes, and the bigger ones become females. Most, however—including *Lycopodium*—produce only one kind of spore, which grows into a bisexual gametophyte. By this, I mean that the little plant has two kinds of sex organs on its body. The technical term for this AC\DC condi-tion is "hermaphrodite," as in Hermes and Aphrodite. When con-ditions are damp enough and the season is right, the sperm swim out of their respective male organs in search of female organs and their eggs. Although some of these gametophytes may be bisexual, they draw the line at incest. They generally avoid fertilizing themselves, although the precise mechanisms for this are not known. Chemical incompatibility between one's own sperm and eggs may be the cause, or it may simply be that the male part of the body matures and releases sex cells before the female part.

Now suppose that a princess pine sperm finds a receptive egg to fertilize. The result of that union will begin to grow while still attached to the parent gametophyte. It puts down some roots and sends up a shoot that develops into a leafy princess pine. Thus proceeds a cyclical alternation of plant generations that goes on and on, from leafy princess pine to gametophyte to princess pine and back again, for as long as the species endures.

In princess pines, kids end up being more like their grandparents than like their parents. Come to think of it, that is not so different from what has happened in some human families I know. Grandparents and their grandkids; that is what you find out there in the woods when you visit the princess pines.

Winter

Winter Woods

It may seem strange, but I often spend more time tromping around in the Northern Forest during the winter than during the summer. For me, summer is a time to paddle around on lakes; the bogs, underbrush, and biting flies can sometimes make bushwhacking pretty miserable then. During winter, many of the birds are gone and some of the mammals are hibernating, but it is easier to get around when the foliage is not cluttering everything up and

when the wetlands are frozen. With fresh snow on the ground, I can often get a better sense of the animal lives going on around me than I can during warmer months.

For one thing, when there is snow on the ground, animals leave tracks. Because our noses are practically useless for anything other than hanging eyeglasses on, we humans often do not know that a creature is in our vicinity unless we actually see it. Winter tracks tell a great deal about animal goings-on for hours or even days after the fact, and reading them can be like leafing through this week's back issues of the local daily paper. In winter, it is easy for the most nasally challenged human busybody to tell who is in residence and what they have been up to. One only needs to read the latest trails and droppings.

Take this morning, for instance. It dawned clear, bright, and windless, as if the curtain of night had been pulled away to reveal a sparkling new Perfect Winter Morning. Days like that draw me outdoors as irresistibly as a hot summer day at the shore. I geared up in my lightest parka, hat, and gloves, put on the sunglasses purchased years earlier for work in the East African desert, and headed into the woods across the road from campus. A nice, groomed ski trail circles Heron Marsh, and I like to patrol it at different times of year to watch the seasonal changes. Winter is when I usually see the most animal signs, and this morning was no exception.

Where the path crossed the marsh outlet at Shingle Mill Falls, a series of fresh skid marks on the snow-dusted ice caught my attention. A river otter had scampered and slid across the frozen surface of the marsh, rushing straight at the open water that tumbles over the dam. A last long slide led to the end of the ice where the otter had tobogganed right over the falls. The snow script was so clear that I could almost see the thick, ropy tail disappear into the rapids.

Farther along the trail, where hemlocks shadowed the ski ruts, I knelt to look into the facing mouths of a tiny snow tunnel that was bisected by the last skier. A mouse had scampered from one hole to the other across the ski trail, probably fearing for its life in the brief moment that its hunching brown back lay open to predators. Tiny paw prints and a light tail drag between them told the story.

Among the thick spruce and fir saplings beyond the hemlocks, the powder was pocked with the signs of a bustling community of snowshoe hares, none of whom I so much as glimpsed directly. Had I pushed my pace a bit faster, I might have spotted several of the splay-footed grouse that left strings of prints entwined in the snowy undergrowth. A coyote had come out of these same woods shortly before I arrived, walking the packed trail for a few yards before bounding back into the drifts. If the ski trail itself had not been so inviting, I could have followed the coyote tracks to try to discover what it had been seeking: perhaps hares, or grouse, or some other private aspect of coyote affairs.

By the time I got home an hour later, I had not seen a living thing other than trees, shrubs, and chickadees *(Parus atricapillus)*. But, thanks to the snow, I felt a greater animal presence on that path than I ever have during the snowless months.

If you enjoy watching whitetailed deer, a good place to look for them is among dense stands of spruce or northern white cedar in low, wet areas. In late fall, the deer haunt these dense conifer groves to find food and shelter and to avoid human hunters who do not like to get their feet wet. In winter, they come for browse and protection from the elements. When deer hang out in a sheltered area such as that, people say they are "yarding up," and you can often find flattened patches of snow where they slept among the sheltering trees. You can find well-packed deer trails in and around such thick conifer stands all winter long, because the animals tend to stick to established paths rather than expend the energy necessary to break new trails. Another reason why deer seek shelter in cedar and spruce groves is that the thick evergreen boughs hold a lot of snow off the ground, so the animals do not have to work so hard to get around.

Still, in my neck of the woods, it is not uncommon to find the tracks of a lone deer in just about any patch of forest. When the powder is deep, the animal's flanks push the snow aside like a plow and its cloven hooves sink through the snowpack, leaving deep prints that slant downwards in the direction of travel. You can tell from such a trail that deer have a tough time moving around much at that time of year, and it makes you wonder what drew that particular animal out of the main trail system.

As the leaves fall and winter approaches, the bucks go into rut. In their ardor to challenge competitors and to attract mates, they often advertise their presence with scents and signs. Even in the depth of winter, you can often tell where a rutting autumn buck once staked his territorial claim by watching for places where antlers rubbed vigorously against the lower stems of saplings. Some of the most popular species to rub against in the north country are maples, beech, poplar, and eastern hemlock. The bark gets scraped off or shredded into strips and the pale, bare wood often shows up plainly against the surrounding greys and browns of trunks and stems. Hunters look for these rubbing-spots, and some say that larger bucks make larger marks than do smaller bucks. Perhaps the bucks themselves interpret the size of a rub that way, too?

I like to try to figure out what the deer are eating by following their tracks through the woods and looking for scrapes in the snow. When the snow is powdery and easily moved by the wind, a strong gale will often scoop out bowl-shaped depressions around the bases of tree trunks. Hungry deer can paw away what is left of the snow in these spots and get at whatever plant food lies beneath. I once followed a deer's tracks from tree to tree in late March, when the blowouts had melted open and exposed sparse tufts of withered grass, ferns, and mosses. Most of the grass tufts were reduced to stubble.

You can also monitor local deer behavior by watching for young saplings and shrubs that have had their budding tips nibbled off. You have to examine the nibbles carefully, though, because snow-shoe hares browse on many of the same shoots and shrubs. Hares generally bite the buds clean off, leaving a smooth cut aligned at about a forty-five degree angle. But because deer lack upper incisors (what you might call "buck teeth"), they must rip each bud loose with their molars or pin it between their lower front teeth and upper gums and yank it off. Deer browsing tends to leave ragged-looking stubs behind.

Older nibblings from previous years have a darker, healed-over look to them. It is interesting to note how high above ground level the cuts are in spring or summer, because it gives you some idea of how thickly the snow can lay in these woods in winter. Many of the sapling tips brushing your waist in June were originally chewed down to snow level.

You may notice that the white cedars that line the shores of many of our lakes often have their lowermost branches pruned off. Each tree is neatly trimmed to the same level, as if by a master landscaper. That is usually where winter-hungry deer chewed off every green thing within reach. If they can get them, deer generally prefer forage such as maple buds and other nutritious hardwood fodder, but bitter-tasting conifer twigs often fill their bellies in the cold months.

Some other animals actually relish the resinous bark and twigs of conifers. When you find bark chewed away from the upper branches and trunks of pines or of other conifers, it is often the work of porcupines. Their dining sites are easily spotted as they ooze great quantities of sticky resin that whitens the tree trunk when it hardens out in the elements. If the snow is not too deep and you happen upon a cluster of large glacial boulders out in the woods, you can often peer into the dark crannies beneath and between the rocks where porcupines like to den. The surest sign of porcupine presence, other than the creature itself, is a large mound of brown, fingerlike droppings sprinkled with loose quills.

As winter draws to a close, you often find the tips of twigs lying all over the snow under conifers, especially spruce, balsam fir, and eastern hemlock. Even after the snow is gone, the ground can be littered with them. Each twig is usually no more than three or four inches long and has been cut cleanly from the branch. Presumably, these cuttings are taken from the upper parts of the tree, because I have never been able to find tipless twigs within reach of the ground.

I have long wondered about those conifer tips. When I first encountered them on the Heron Marsh Trail, I thought that they had simply broken off during storms. But such fresh, healthy twigs are tough, and many of those scattered on the snow were cut off in mid-twig rather than at the potentially weaker branch nodes. Some animal must have been doing it. By the process of elimination, I narrowed the options down to red squirrels, but I have yet to catch an animal in the act. It is not even clear why such behavior should be worth the trouble. Conifers do not have much sweet sap worth releasing in this way, and even if they did, their bitter, gluey resin would probably spoil it. None of the tips that I have found show any signs of grazing damage other than the cut ends themselves.

A chance conversation finally led to what may be the solution to the Conifer Tip Mystery. Some aquaintances from Kansas visited recently, and in the course of conversation they mentioned that they watch squirrels in their back yard nipping leafy branch tips from maple trees. These squirrels let large numbers of maple tips fall to the ground, then they scamper down and carry a select few of them away for nesting material. The rest is left to litter the forest floor. I wonder if our red squirrels do the same thing with Adirondack conifers. They mate in February and March, and the first batches of young (red squirrels generally produce two litters per year) are born in March or April, so the timing is right. If you get to look at a red squirrel nest up close before I do, I hope that you will get in touch to tell me if it is full of fresh green conifer tips.

If you enjoy studying plants that do not have flowers on them, you do not have to give up botanizing in winter. It is not very difficult to identify leafless trees by examining their trunks, because each species has distinctive colors and textures. If the branch tips are within reach, you can clinch your identification with a look at the buds. Conifers are among the easiest to recognize because of their distinctive evergreen needles that vary in length, shape, and cluster-size. Birch trees have bright white or yellowish papery peeling bark when fully grown and smooth, copper-colored bark covered with white flecks when they are still saplings. Beech trees are easy to identify, too, because of their gray bark that looks like smooth skin stretched tightly over muscular limbs.

Around populated areas you can almost bet that if it is a beech tree, it is going to have someone's initials carved in the trunk, because some people cannot resist scarring such smooth gray surfaces. It probably has something to do with the similar inability of some of us to resist walking across an untrammeled expanse of fresh snow. When you get farther beyond the reach of people, you can find more interesting things in or on beech tree trunks. Insects bore holes in them, woodpeckers go after the insects, fungi and mosses sprout in or around the holes, the remnants of the last blowing snow squall may cling to the windward side of the trunk, and algae often stain moist areas of the trunks. I some-times pause to examine beech trunks that are tinted with green algal coatings, looking for signs of grazing. I almost always find pale, meandering trails, a fraction of an inch wide, squiggled through the green. Perhaps it is the work of slugs, though the trails that you see in winter presumably would therefore have been made in earlier, warmer months.

I have also found more delicate grazing-traces on algae-coated white birch bark. The finely filigreed scrapings are tightly packed and each individual mark is about one millimeter thick. The in-tricate patterns remind me of ornate tatoos on the skins of South Pacific islanders, and I would love to know what makes them. Insects are a likely possibility, though, again, slugs may be the true artists.

I noticed something new but rather depressing about beech tree bark on a recent back-country ski trip. Most of the beeches in my part of the Adirondacks have what appears to be a white dusting of lint all over their trunks, as if they lived just down-wind from a laundromat. A forester friend explained that it is an insect-borne fungal infection that spreads through the wood and bark of beeches and eventually kills them. Who knows what tomorrow's north woods will look like, as the comings and go-ings of disease, natural forest succession and competition, climate change, and air pollution rearrange the ranks of the forest trees?

One of my favorite things to look for in an Adirondack beech grove is bear sign. The next time that you walk in a beech stand in the winter (hopefully before the fungus gets them all), look up into the naked treetops and see if you can spot broken branches that still have last fall's brown leaves attached to them. Quite often, that is bear damage. When the beechnuts ripen in the fall, black bears like to sit up in the canopy and pull branches down to them, like fat Caesars amid dangling bunches of grapes, so they can chew mouthfuls of beechnuts right off the twigs. Those broken branches do not seem to get the normal plant hormonal signals to drop their leaves when the others do, so they are easy to spot in winter.

When you find a bear tree like that, walk up to where you can get a good look at the trunk. You may discover bear claw marks going all the way up to the top and back down again. This fall's claw marks may still have fresh wood and pale brown bark cortex exposed. Older marks are black and scabby. You can see where each paw went, and it is easy to imagine that bear shimmying up the trunk with its claws splayed out and digging into the dense wood. You can even get an idea of how big the bear was by the size of the claw marks; the broader the spread of the claws, the bigger the bear.

With some careful observation and a little imagination, a winter walk in your local forest can leave you feeling like you have touched base with the neighbors, even without seeing any of them face to face. Once you get used to noticing tracks and other signs of forest life, you never have to feel lonesome out there on the cold, snowy trails of winter.

Northern Lights

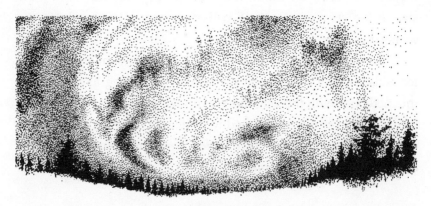

A spectacular display of northern lights appeared over the Adirondacks early on a November evening in 1992. The sky blazed with curtains and flashes of green light, and bright red jets lanced the green. A friend came out to watch them with me in the midst of a large, open field. I stood open-mouthed, staring upwards and shivering with a mix of excitement and the night's chill as he spoke softly, as if to the sky itself. He was explaining the light show as best he could.

Nearly every time I watch the northern lights with other people, it seems, the same conversation begins. "What causes it?" "I've heard it's from sunlight reflecting off the polar ice caps." "Listen; do you hear the whispering sound? It must be from the hissing of the fires up there." These explanations are often delivered with great conviction in hushed tones that suit the moment's atmosphere of awe and mystery, and I normally do not have the heart to question them out there beneath the flashing sky.

There is a long history of interesting ideas about what causes the night sky to ignite. Vikings said that the flames-in-the-heavens were flashes from Vulcan's forge. Inuit peoples of the far north said that they were caused by ancestor spirits. Early scientists suggested that auroras were refracted images of sunlight curving in from over the pole; their shimmering was thought to result from air movements, like those that make mirages ripple and shift. Nowadays, the standard scientific explanation is that the lights occur when solar radiation plows into the upper atmosphere and sets charged molecules aglow until the sky lights up like a fluorescent lamp. More on that later.

The technical name for the northern lights is *aurora borealis,* translating roughly to "northern dawn." The term originated with a seventeenth-century investigator named Gassendi, who fused the name of the Roman goddess of dawn, Aurora, with that of the god of the north wind, Boreas. For those of you who are a bit rusty on ancient mythology, Aurora's mother was the moon, and her father was the sun. But in spite of her impressive pedigree she never managed to impress Orion, the male of her dreams and the model for a familiar constellation. This proved so upsetting to Aurora that to this day, she weeps bitter tears that fall to Earth as frost.

Not all auroras are classified as "borealis," however. You can also see them above the south pole, though they are less common there. The "southern lights" are called *aurora australis.* Their existence was hypothesized in the early 1700s but not confirmed until Captain Cook circled Antarctica in 1773.

One of my most awesome encounters with the northern lights took place on a June evening on the coast of Maine, near Damariscotta. I was teaching marine biology there, at the Audubon Society Ecology Camp on Hog Island. The campers were all abed, and some of us off-duty staff crept quietly out to the dock for a moonlight row around the harbor. The Milky Way spilled across the silent sky and glistened on the dark mirror of the bay as we launched our creaky wooden skiff. The instant I dipped the oars into the water, a cry went up from my companions. Pools of light welled where the blades had swirled, and a river of greenish phosphorescence trailed from our stern. It was a luminous sea, not unusual on that coast but especially intense that evening, ignited by a thick bloom of bioluminescent dinoflagellate algae and comb jellies. The reflections of a million stars mingled with submerged galaxies of plankton sparks.

"All we need now," I mused, "is the northern lights." As if on cue, what appeared to be a noctilucent cloud above the northern horizon shifted suddenly to the right, then trailed a ghostly finger across the sky. Starlight poked through what a moment before had seemed opaque. Within half an hour, the northern half of the sky was aflame. Feeling too overwhelmed to row back and wake the campers to share in what by now felt like a private gift from the universe, we basked in the three-dimensional glory of it all.

I find the current scientific explanation of how auroras form to be fascinating, but it is rather involved and therefore not very convenient for presenting to friends out in a rowboat on a phosphorescent sea or while shivering in a dusky field. To get comfortable with it takes some careful thought, a good imagination, and a willingness to tackle a bit of technical jargon. On that note, here is a summary of what I was able to dig out of the recent scientific literature. I promise that it will not be nearly as technical and complex as what I had to wade through.

Auroras form only around polar regions. That is not to say that they only form around our own; they have also been photographed hovering like luminous halos over the poles of Venus, Mars, Jupiter, and Uranus. There is a good reason why the poles of planets develop these light shows, and it has to do with magnetism, electricity, and something called the *solar wind*.

I will start with the solar wind first. The sun is a massive, runaway fusion meltdown, an ancient, ongoing nuclear explosion. It throws tremendous quantities of energy and of subatomic particles in every direction, far out into space. Our planet, like the others in our solar system, sails through a storm of solar ejecta, visible light, and other radiation. Sometimes the storm waxes and wanes as sporadic jetlike flares and gigantic bubbles of matter and energy called CMEs (coronal mass ejections) lift off the surface of the sun and fly out into the void. Much of the torrent consists of negatively charged electrons and other charged particles. This flood of particles is the solar wind.

The solar wind normally contains about five flying particles per cubic centimeter and is like a barrage of subatomic projectiles roaring out of the sun at tremendous speeds (up to one thousand kilometers per second). Although the particles are smaller than atoms, they do exert a force on things with which they collide. The first direct evidence of solar wind was reported in 1931, when it was noted that the tails of passing comets always point away from the sun no matter what direction the comet is moving in. It is solar wind that blows sparkling ice and dust off the surfaces of comets to form those bright, trailing tails.

Magnetic fields make up the other half of the story. The earth generates magnetic energy through poorly understood events in its molten depths. As a result, our whole planet is surrounded by invisible magnetic field lines that loop from the south pole to the north pole like those sweeping around the tips of a bar magnet.

Actually, the shape of our magnetic field is heavily distorted by gusts of the solar wind and from the sun's own enormous magnetic field. But the details of its shifting shapes are quite complicated and I do not think it is worth trying to explain them here (largely because I do not really understand them myself). What is most important to know for our purposes is that when the charged particles of the solar wind meet the earth's billowing magnetic field lines, many of these charged particles change course to follow those lines in toward the poles.

What happens then, it turns out, really is much like a fluorescent lamp. Such lamps work by sending electricity (that is, a stream of electrons) through a tube full of excitable gases that glow in the presence of an electrical jolt. The upper atmosphere also contains lively gases, because x-rays and ultraviolet rays from the sun turn normal air molecules into excitable charged particles, or ions; that is why it is called the *ionosphere*. The ionosphere lies between about sixty and three hundred kilometers above us, and it acts something like a lamp tube full of excitable gas. The earth's magnetic field lines guide electrons from the solar wind into the ionosphere, just as flickering magnetic fields direct electron beams onto your TV screen or as wires lead electricity into a light bulb. When the electron streams meet the gases, the gases light up. The glow itself comes from the energy exploding out of gas molecules that have been struck by flying electrons. Small fragments fly out from the molecules as well, forming still more charged particles in the ionosphere. Some of the energy emerges as visible light, and if enough gas molecules do this at the same time, you can see the glow from the ground. That is when you see an aurora.

Most auroras are pale whitish or greenish in color, the result of energy flashes given off by countless oxygen molecules. The reds and pinks that you see at the bottoms of auroral curtains are given off by nitrogen molecules. You also tend to see more reds overall when auroras extend closer to the equator than usual, sometimes even into the southern United States. In those instances, much of the red comes from oxygen atoms that have been energized more than usual by the extrapowerful magnetic disturbances that drove the auroras to such low latitudes in the first place.

But not all of the energy given off by these gases is visible. The shattering of air molecules and ions also emits x-ray, ultraviolet, infrared, and other invisible forms of radiation as well. Satellites have picked up x-ray and ultraviolet images of auroras that are otherwise invisible to the naked eye. Such satellites have detected auroras forming over the daylit half of the north pole, where we cannot see them because the daylit sky obscures the auroral light coming down from above (and because most of us don't live at the north pole). This is not unusual; in fact, there is often more auroral activity over the daylit side of the pole than over the night side.

Perhaps the most spectacular aspect of the northern lights is the long, rippling, ray-streaked curtains that they spread across the northern half of the sky. It is difficult to get a sense of how big those curtains are, because there is nothing up there to measure them against. But they are enormous. The curtains usually start as diffuse, ragged edges hundreds of kilometers above the earth and bottom out in thick, bright hemlines at roughly one hundred kilometers altitude. The increasing brightness with proximity to Earth results from the natural downward thickening of the atmosphere; because there is more gas at the bottom, there is more to glow. The bright bottoms of the curtains are so distinct because they mark a boundary in the atmosphere below which the air is too dense for the solar electrons to penetrate. Most curtains are less than one kilometer thick, because of the relatively thin, sheetlike nature of the magnetically channeled solar electron streams, but they often extend for thousands of kilometers to the east and west. Within the curtains themselves, you may often see streaks of light flashing upwards. These are tornadolike vortices of whirling solar wind electrons driven by turbulence in the electromagnetic field.

Sometimes an auroral display suddenly intensifies shortly before midnight, as if some giant hand turned up the brightness knob. The extrabright curtains drift polewards towards an expanding zone of light-formations, rippling majestically from west to east as they go. Some of these great ripples, while appearing to move slowly from our distant observation posts, actually move through the curtains at rates close to two thousand miles per hour. After about an hour, the display subsides into a dim fog that can cover the entire polar region. These events seem to result from complex disruptions in the earth's own magnetic field structure (polar magnetic substorms), which occur four or five times a day but are not normally noticed here on the ground unless auroras are present.

If you were to look at an aurora from space, it would resemble a winding ribbon of light curled in a halolike ring one hundred miles or more above the polar regions. In fact, there is an oval ring of energized gases and electromagnetic activity centered over each pole all the time, but it does not always glow enough to produce what we would normally call an aurora. The auroral oval ring gets brighter and dimmer, larger and smaller with fluctuations in solar wind and magnetic field conditions; when it is bigger, you can see auroras at lower latitudes than usual. And sometimes the diffuse glow of ionized gases leaks out of the borders of the oval and spreads a haze over the pole. You can often see similarly dim patches of auroral haze at lower latitudes as well.

A good way to tell auroral fogs from real fogs or clouds is to look for stars in the midst of them. If you can see stars through it, it is probably an aurora. To clinch your decision, watch one spot very intently for a minute or so, and see if you can catch it moving or changing shape. If you do, you know Vulcan is working late at the forge.

As a general rule of thumb, I would simply say that you might as well check the night sky on any clear night of the year, at any time before midnight. However, the best time to look for the northern lights is often around the spring and fall equinoxes, when the earth is most directly exposed to the stormier high latitudes of the sun. Other good times for aurora-watching are thought to occur when there are more sunspots on the sun than usual, every eleven years or so. Sunspots are mysterious dark blemishes on the sun's surface where temperatures are less intense than normal, but the sun's magnetic field is strongest.

The reason why sunspots and auroras often develop at the same time was discovered only recently. Gigantic, bubble-shaped CME eruptions often develop in the sun's flaming rim (the *corona*, Latin for "crown") at the same time as do sunspots, and these eruptions throw incredibly powerful bursts of solar wind far out into space. A single CME represents as much energy output as about five billion power stations. For all their size and violence, CMEs are visible only in photos taken by space probes that fly free of our hazy atmophere. The cameras involved have to be supplied with devices that block out the brilliant face of the sun, in much the same way that the moon does in a total solar eclipse; this makes the corona more visible. Not surprisingly, nobody knew about these eruptions until the early 1970s, but it soon became apparent that they are the principal sources of solar wind and magnetic disturbance that create the earth's strongest auroras. When the most violent solar outbursts strike our atmosphere, the resultant electromagnetic interactions can cause more than auroras here on Earth. They may wreak havoc with compasses and radios, send surges through power lines, and generate electrical currents along oil pipelines.

Before the advent of space probes, scientists assumed that solar flares, the bright eruptions associated with sunspots, were the main driving force behind auroras. They are much smaller than CMEs but are readily visible from Earth and therefore have been known about for quite a while. Even now, you may hear about solar flares on the evening news when auroras are sighted. The supposed connection between sunspots, solar flares, and auroras dates back to the late 1800s, when several centuries worth of observations were compiled to show that they all seemed to occur at the same time. It seemed only natural back then to assume that this correlation also proved a causal relationship. Now that CMEs are the rage among astronomers, newly enlightened scientists have reviewed the old astronomical records that were once used to show that solar flares cause auroras. They now report finding only a slight correlation between flares and auroras in the data, with little evidence for a causal connection between them!

Individual solar storms can last for days or weeks, so the same disturbance can generate several nights' worth of auroras. It will not last for more than one or two nights at a time, though, because the sun spins just like the earth does, and the paths of its coronal explosions will point away from us as the sun spins on its axis. You have to wait about twenty-seven days, until the sun spins all the way around, before you know if that particular storm is still active. If it is, then you may be in for another show.

This is not meant to imply that the northern lights pay much heed to predictions. I have seen more of them in the full heat of summer than in spring or fall, although a certain reluctance to face Adirondack night temperatures in March or October may play a role in my case.

Many people report hearing whispering sounds coming from the northern lights. Simply put, there is no scientific basis for such a phenomenon. The auroras are a long way away from you, usually between sixty and six hundred miles overhead and reaching much farther than that at the far ends of the curtains. It is pretty unlikely that you could pick up sounds from such thin air over such huge distances. I suspect that what people really hear, if they hear whispering in the distance, is distant breezes rustling through treetops far off in the darkness.

Still, it is difficult to convince "ear-witnesses" of that. I guess it just makes too much gut-level sense that such an enormous, ghostly display should have an ethereal sound track to go with it and, perhaps, a hushed audience whispering softly in the darkness.

Bird Feeder Biology

More than sixty-five million Americans maintain bird feeders, dishing out about one billion pounds of seed each year. This puts bird feeding right up there along with gardening among America's most popular nature-oriented activities, and it classifies feeders as significant wildlife management tools. Birdfeeding is also popular in Europe; I remember my local newspaper in southern Sweden (where I lived and worked in 1987) reminding folks to keep their feeders filled to help the birds through a particularly severe winter. I also remember my Swedish acquaintances blaming me for bringing with me the worst winter weather in living memory. I made sure my feeder was overflowing at all times, to help atone for my alleged sin.

When things get rough out there in the winter landscape, it may seem obvious that bird feeders can play a vital role in bird survival. In fact, it is not at all clear that this is so, for—believe it or not—relatively little scientific study has been done on the precise role that feeders play in bird ecology and behavior.

Because I would like to limit this topic to birds that you are very likely to encounter at your own feeder, I will focus on black-capped chickadees (*Parus atricapillus*). Being fairly small birds (they weigh only about four ounces apiece), chickadees have high metabolic rates that make up for the rapid loss of heat from their tiny bodies. Researchers at the University of Vermont report that during long winter nights, chickadees lower their body temperatures by as much as eighteen degrees Fahrenheit and keep their muscular heat generators running by shivering. To fuel those inner furnaces, they burn the equivalent of a little more than one gram of body fat per day. This means that at least in winter, they must eat every day or they may run out of food energy and die. Chickadees often do not have the luxury of sitting out a storm in some sheltered nook, although there is evidence that the little birds can fall into a sort of deep sleep or torpor to conserve energy when conditions are just too harsh for foraging.

In summer, chickadees live mainly on animal foods, especially moths, butterflies, and spiders. They also eat other insects, small snails and slugs, and whole flocks have even been seen picking fat out of deer carcasses like little vultures. But during the colder months, animal food is harder to come by and the chickadees switch to a roughly fifty-fifty mix of animal and plant foods. During the autumn months, they eat berries from the likes of bayberry, poison ivy, and blueberries and pluck seeds from goldenrod, ragweed, staghorn sumac, and other plants. In winter they rely most heavily on the seeds of cone-bearing evergreens, especially eastern hemlock. They eat the sweet frozen sap from the broken tips of sugar maple branches when the sap is running and scrounge among the furrows on tree trunks for dormant spiders and insects and their cocoons.

One strategy that chickadees use to hedge their bets for winter survival is the storing of extra bits of food over a wide area during the late autumn. Seeds are the most commonly stored items; if you start your feeder in November, you may be able to watch your chickadees remove the husks from individual sunflower seeds and slip them into knotholes and crannies on the undersides of branches, under the loose, peeling bark of birches, or even on the ground. Sometimes insects have their heads nipped off and the rest of their bodies stored like seeds as well. Amazingly, the birds seem to be able to remember where most of these caches are for at least a month, and probably longer.

Another way for a chickadee to increase its chances of survival is to team up with other chickadees during winter and to share the work of finding food. When a chickadee arrives at your feeder early in the morning, chances are good that several more are also about to land. White-breasted nuthatches *(Sitta carolinensis)* and downy woodpeckers *(Dendrocopos pubescens)* often tag along with the roving chickadee teams, depending upon their sharp eyes and inquisitive natures to ferret out food that often ends up being shared by all. Other birds may cue in to the wisdom of following hungry chickadees in winter, as well. Some of my students repeatedly played recordings of various bird calls beside a bird feeder near the campus and found that redpolls *(Carduelis flammea)* were significantly more likely to show up when chickadee calls were played than when the calls of other birds were broadcast.

Working the woods in groups may have the added advantage of providing warm snuggling partners that can help you survive frigid winter nights. Ruby-crowned kinglets *(Regulus calendula)* have been shown to do this, but nobody knows if chickadees do it too. Individual chickadees and nuthatches are also known to conserve body heat by holing up in hollow trees and pre-existing trunk cavities that were dug by woodpeckers.

Knowing how chickadees struggle to find food in winter may make you feel obliged to do all you can to help them by putting out feeding stations for them. But there is very little information available about how important your feeder really is to wild birds. Some of the few published research reports seem to indicate that chickadees generally prefer to cover a wide area in their foraging, even when reliable feeding stations are available.

One study conducted in Wisconsin between 1983 and 1985 involved feeding 348 wild chickadees (all of them banded) on black oil sunflower seeds all winter. The researchers estimated that the average bird took thirty to forty seeds per day regardless of weather conditions, traveled less than one-half kilometer from the center of its home range to get to the feeder, arrived three hours after sunrise, and departed two hours before sunset. But, most surprisingly, they also found that these birds got only about one-fifth of their daily food energy requirements from the feeders. The bulk of their nutrition came from the usual chickadee winter fare of wild seeds and tree-hugging insects. Why the birds did not stay close to the full feeders remains a mystery, although the researchers did have a reasonable explanation to offer.

Normally, wild birds have to forage widely in winter to find enough food, and they can never be sure that any given location will yield food reliably. Because the consequences of missing a meal in winter are so serious, it might be in the best interest of chickadees to stay in the habit of moving on to other potential feeding sites in order to avoid the risk of running out of food at any given location some day. Recent studies suggest that some chickadees do have a marked preference for low-risk foods that vary little in abundance. I suppose this means that you ought to keep your feeder well filled at all times, if you fill it at all, but that this might be more of an act of politeness than of life support for the birds.

Those of you who do keep the feeder filled are probably well aware that birds are not the only visitors. Red squirrels *(Tamiasciurus hudsonicus)* are notorious feeder raiders in the Northern Forest region, and they tend to be as good at figuring out ways to reach birdseed as we are at inventing ways to discourage them. Perhaps the best way to deal with red squirrels is to learn to enjoy them as much as we do the birds, although there can certainly be problems if the squirrels decide to treat the birds like another course in the banquet (I recently found a report of a red squirrel eating an adult male goldfinch, *Carduelis tristis*). Whatever your feelings about them may be, squirrels are certainly entertaining to watch, and before long you can begin to recognize individuals by their distinctive sizes, color variations, scars, and so on.

But if you decide to feed red squirrels in addition to birds, you should be prepared to face the fact that you may be altering their breeding and territorial behaviors as well as their feeding habits. This might also be true of birds to some extent, but squirrels are much more restricted in their wanderings than are birds, and a large food source within their territory can have much more impact on them. A study in Alberta showed that providing wild red squirrels with an unlimited supply of sunflower seeds made them heavier at the start of the late winter breeding season (March or April, in the Adirondacks) than they would normally have been. This, in turn, made the females go into heat about three weeks earlier than usual. Juveniles born earlier in the season are more likely to win territorial contests than are late-borns, presumably because of their slightly greater size and experience. And another study found a two- to four-fold increase in local red squirrel populations resulting from the arrival of youngsters from elsewhere during their autumn wanderings. As soon as the food supply was cut off, the squirrels dispersed. The point of all this is that although wildlife may not actually depend upon your feeder for survival, the impacts of your generosity may nonetheless be greater than you anticipate, at least in the case of red squirrels.

Another mystery that is fun to investigate at home is the question of what makes the best all-around bird food. Store-keepers are more than happy to sell you expensive thistle seeds (*Cirsium* species) or grab-bag mixtures of assorted seeds that most birds just pick through to get at their favorite items. If you are one of the millions of Americans who feed birds, you probably already have some idea of what your birds go for, but I encourage you to experiment as widely as possible before you settle on that perfect formula. Try table scraps, dog food, or whatever leftovers come to mind. Maybe the goldenrod and ragweed in a vacant lot could provide you with enough free seeds in autumn to save you a pile of money.

In my part of the woods, about the only thing the birds seem to be interested in eating is black oil sunflower seeds. I have even had raccoons stop by on spring evenings to scrounge them out of the day's leftover husks. For some reason, the black oil seeds are held in much higher esteem here than are the larger striped kinds, although I cannot tell you why. I just go with it and buy black oil seeds by the fifty-pound sack all winter long.

Several of my students have looked into this matter of seed preference in the course of their biology laboratory research assignments, but their work has yielded more questions than answers. One group of students painted black oil seeds to look like striped seeds and put them out in dishes along with normal sunflower seeds of both varieties to see what happened. The chickadees went for the black oil seeds regardless of whether they were painted or plain, a result that suggests that it is not simply color patterns that sparks this preference. Do Adirondack chickadees prefer black oil seeds simply because they are easier to open than the big striped ones? Students put out dishes of shelled and unshelled seeds of both types, and the chickadees still went for the black oils complete with shells. That result still does not make sense to me; I would have assumed that the birds would choose to save energy by eating the pre-shelled seeds.

That idea of saving energy is a central theme in many biological field studies. Animals are not supposed to do anything that wastes energy, presumably because they are involved in an ancient struggle for existence that does not permit the squandering of time or of energy. Then why the preference for shelling their own seeds? Even stranger was the result of yet another student project that looked at individual black oil seeds.

Look closely at the seeds in your feed sack and you will notice that they vary somewhat in size. Some are also empty, the work of insects that ate the innards and then burrowed out through tiny holes in the shells. I have watched nuthatches land on my feeder with a businesslike air and sort through the day's offering by flinging away seeds that, I suspect, have those little holes in them. Because empties would not be worth opening, the nuthatch behavior probably fits neatly into the aforementioned model of energy conservation. But there are also perfectly good seeds that are, nonetheless, significantly smaller than the rest. One of my students wondered if local birds would preferentially seek out the larger seeds that yield more food when opened. We were both surprised to discover that there was no such preference. As long as there was "meat" inside the seed, and it was a black oil seed, the chickades treated it like any other.

These results seem to challenge published studies that suggest that chickadees tend to save energy by sticking with "sure bet" food sources rather than variable ones. One such study gave wild birds a choice of two feeding stations to come to. The same numbers of seeds were available at either station, but their distributions were different. One station was a standing tree with holes drilled in it and the holes packed with equal numbers of seeds. The other was similar, except that some of the holes contained fewer seeds than others. The chickadees in that study chose the station with the predictable seed numbers.

But other studies find that chickadee behavior varies from time to time and place to place. That makes more sense to me; why would an intelligent animal limit itself solely to simple, stereo-typed behaviors in a complex, changing environment?

Yet another aspect of feeder research is keeping track of who shows up every year. I have yet to get the same mix of birds at my feeder each winter. The first year I started putting out seeds, my place was mobbed by pine siskins *(Carduelis pinus)*. The next year there were none, and I feared that they had all been eradi-cated by some mysterious siskin virus. In fact, some friends who live farther south assured me that they had all moved down to Albany that winter.

Once a grackle *(Quiscalus quiscula)* stayed on too long for some mysterious reason. It was December, and the poor thing's com-panions had long since headed out. It sat hunched miserably among the chickadees, ineptly opening seeds and looking weaker every day. I soon noticed that its left foot was encased in a ball of ice. There was nothing that I could do to help, because the grackle flew off whenever I stepped outside, but the drama was wrench-ing. It never warmed up enough to melt that ball of ice. Finally, just after Christmas, I found the frozen body of the grackle lying in the snow beside the feeder.

During the frigid winter of 1993–94, I had redpolls at my feeder for the first time in four years. They just showed up, by the dozen, one crisp day in January. But stranger still was the total absence of blue jays *(Cyanocitta cristata)*. There are almost always blue jays around, but not that winter. My neighbors and I feared the worst

for the jays, until flocks of them showed up in late April, as noisy
and as obnoxious as ever. The redpolls disappeared then, perhaps
as a result of the jays' return. The question of what drove the jays
away that year is all the more confusing because they have stayed
on through other severe winters.

But whatever the weather, season, or year, I have always been
able to enjoy chickadees. I often wish that it was easier to tell
them apart (without having to risk hurting or encumbering them
with leg-bands or other tagging devices), so I could follow the
fortunes of individual birds. But I cannot even tell males from
females. The chickadees themselves have no trouble telling each
other apart, of course, and you can see this for yourself by watching how they interact at your feeder.

Like many birds, chickadees have a social pecking order in their
flocks, with some birds intimidating or overpowering others when
they come to feed. Low-ranking birds tend to fly off the feeder as
soon as high-ranking individuals arrive and land as if they owned
the place. Everyone seems to recognize everyone else's position in
the local pecking order at a glance, presumably by noting subtle
differences in anatomical features, calls, or distinctive styles of body
movement. Dominant birds, for example, are more likely than low-
rankers to tip their heads up or to make intimidating gargling noises.

Sometimes birds of roughly equivalent rank (or one high-ranker
and one uppity underling) face off when one seeks to displace
another. Researchers at the University of Wisconsin have dubbed
this sort of encounter a *ballet*, because the participants energeti-
cally shift their positions relative to each other as if they were
dancing. The contenders are probably sizing up each other's
strengths and weaknesses as they do this, and they generally
manage to figure out who wins without actually having to fight.
As the end of such a ballet nears, the loser tends to face away
from the winner and the winner faces the loser.

I am sure that all sorts of secret social dealings are carried out around our feeders under the cover of uniform markings. Maybe some day, with careful observations and a little luck (like having an easily recognizable albino chickadee take up residence in your back yard), you will be able to discover some of the intricacies of chickadee life that still elude us.

Snowfleas

 When I came home from work yesterday, the message light on my phone machine was blinking. It was a breathless call from a friend whose house was being assaulted by an army of mysterious little bugs: "They're land-based, but they are OK on water. They look like tiny specks, but then, *ping*, they're gone like fleas. I wonder if you could tell me how I can, um, get rid of them. Besides stomping all over them, I mean. They're trying to get into my door. I don't want them here!"

I called and reassured her that they were nothing more than swarming deadly bloodsucking Adirondack plague fleas. She knew me well enough to know that this meant that they were perfectly harmless. They were not fleas at all, and they did not particularly wish to invade her house. I have seen them swarming against the side of another friend's house in late April, but that was probably only because the house stood in the path of their migration. No, these were not fleas (Order Siphonaptera, translating from the Greek to something like "wingless siphons") but rather spring-tails (order Collembola), and they thought no more of biting anybody than we would of biting them. Nonetheless, these kinds of insects do hop around in a manner so reminiscent of flea-leaps that most people call them snowfleas.

You have probably seen snowfleas (*Hypogastrura nivicola*) at some time or another if you spend much time out in the woods in late winter, but you may not have realized it at the time. They are so small that you are not likely to "see a snowflea." You normally see *many* snowfleas sprinkled across the snow like dirt or soot. When they are really swarming, the snow can turn gray with them.

One warm day in February, I skied a forest trail that was literally black with them. I could not imagine how many lay crushed in the ski ruts. The sickly odor that rose from the crushed bodies reminded me of fish slime. Perhaps the smell was a sort of warning chemical released by springtails in distress, or maybe it was just the stench of mass insect death.

You have to get a close-up view of a springtail if you are going to see why they are not fleas. A flea packs six legs beneath an almost crescent-shaped, jointed brownish body. It stands tall and narrow, and when it walks, it tends to sway like a galleon on the high seas. You never see fleas in dense swarms out in the woods, except perhaps on the miserable hide of some luckless mammal or bird. A typical springtail, on the other hand, is blue-gray and stands a bit closer to the ground. It has a distinct head with a pair of dark eyes (each consisting of sixteen smaller ones) and with floppy, segmented antennae. It even looks kind of cute. Being a wingless insect, it has six legs for walking, but at first glance you might think it had an extra pair of legs folded under its belly.

That is the spring on the springtail's tail. It is forked, leading specialists to call it a *furcula,* after the Latin word for fork. It works under the same principle as do catapults. When the animal wants to move fast and far, it flicks its furcula against the ground and tosses itself high into the air. A springtail is small, only a couple of millimeters long, but it can jump ten to twenty times its own body length at the drop of a furcula. A furcula is a handy device for a springtail to have, because the little beings are pretty defenseless otherwise. Many of them are distasteful to insect predators, and most are covered with loose scales and bristles that make it hard for predators to grab them. But the furcula is the collembolan answer to wings. When the animal is not leaping about, it keeps the furcula clasped primly against its third body segment.

If you take an even closer look at a springtail, you will probably also notice yet another structure affixed to its underbelly, this time to the first body segment. It is a tubelike thing, called a *collophore*, that sticks out of the animal's chest and confuses scientists. At first, naturalists supposed that the collophore was used for sticking a springtail to whatever it happened to be walking on. In fact, that is how springtails got their official scientific name, Collembola: *coll* is Greek for "glue," and *embola* means "bolt or wedge," like the one you have sticking out of your chest if you happen to be a collembolan. The term "collophore" derives from *coll* and *phoros*, and translates as "to bear glue." Unfortunately for the terminology, specialists now suspect that collophores have more to do with drinking than with glue. Should this turn out to be correct, I suppose some earnest young taxonomist will soon suggest changing the official name to "Bibembola."

In fact, some scientists have taken into account the springtail's many oddities and have considered putting it into a class of its own, separate from the insects. They have the requisite six legs, but no other insect has a furcula or a collophore. Furthermore, springtails do not metamorphose from a larval infancy into adulthood, as do most insects. They just hatch out of their respective eggs and get bigger, shedding their armorlike exoskeletons as they grow.

There are many kinds of springtails, probably many more than have been recorded by scientists, and most do not live on snow. I have seen little gray rafts of *Podura aquatica* drifting about on the surfaces of tide pools on the coast of Maine. I have found them clinging to the roots of salt marsh grasses *(Spartina alterniflora)* below the tide line in North Carolina. Those salt marsh spring-tails were covered with a water-repellent white wool that seemed to keep them dry down there in the wet, anoxic mud of the flooded marsh. I suppose that they breathed through the same air tubes that bring oxygen down to the roots of the intertidal grasses. People have reported finding springtails throughout North America and in Europe, so you can probably find them just about anywhere you bend down and look for them.

Most of my personal experience with Collembola, however, has to do with the ones that darken the snow in late winter and early spring, the ones that everyone calls snowfleas. I wish that I could tell you exactly what they are doing out there on the snow, but I cannot. Nor could anybody at the state museum in Albany when I called them, nor at Cornell University's prestigious ento-mology department, nor at the Smithsonian Institution. I even tried launching my own half-baked research program to answer the question for myself. As you will soon see, I had plenty of fun but did not discover what I set out to find.

One day in March, we had a brief thaw, and fresh hordes of snowfleas appeared on top of the snow. I noticed that they col-lected in hollows in the snow, especially in ski tracks and in foot-prints, and that they seemed to favor the shady sides of rutted tracks. Here was a chance to solve the mystery of why snowfleas appear on the snow.

I began by trying to find out where the creatures were coming from. Wondering if the snowfleas were working their way up through the snowpack, I dug into the snow in a thinly populated section of drift, looking for stray dark specks. I found none, other than the occasional dive-bomber from the cliff edges of the hole. I had thought that the congregations gathering inside footprints and ski tracks might have been evidence of migration within the snow itself, because breaking through the surface with boots or skis would probably expose burrowers. Instead, I learned that the tracks just collected animals that jumped into them and had trouble jumping back out.

By looking closely at individuals inside my own footprints, I could see how much trouble some of them were having. I watched one laboring up the rough face of a snow cliff that I had just made with my right foot. It got a few inches up the wall, then seemed to lose patience. It gave a flick of its furcula, presumably to speed itself along. Of course, furcula flicks throw the animal away from whatever surface it is walking on. Because this one happened to be walking on a vertical surface, the result of the flick was to send it sailing out into space and back to the bottom with a silent crash. I felt like I was watching the torment of the damned, as the insects endlessly climbed and tumbled. Within half an hour, the footprint was shadowed with black flecks.

I went back indoors and came out shortly with a freshly brewed mug of coffee. I sat down on the sun-drenched wooden steps leading to my kitchen door and leaned over to watch a swarm on a glistening snowbank. The bright sunlight poured its energizing warmth into my winter-pale arms and face, making it very easy to stay put for a long look at the wealth of life around me.

Placing a clear sheet of plastic over a patch of snow, I counted the snowfleas under it. There were as many as twenty per square inch of snow, but they can be much more densely packed than that. A book on my shelf claims there can be one hundred thousand of them in a cubic meter of forest soil, and still more when they are migrating. More on that later.

I set my coffee down and leaned closer to peer through a hand lens in hopes of seeing what the snowfleas were doing down there on the melting snowscape. The ones on the snow itself were crawling and hopping about with no purpose that I could discern. The ones on the part of the steps soaked by snowmelt sat still with their faces pressed against the wood. The ones on the dry parts of the steps milled about like their compatriots out on the snow. Then I noticed that the wooden steps were slippery with something, probably algae or fungi. It looked as though the snowfleas were drinking or grazing on something on the steps.

Maybe the snowfleas were out on the snow in search of food? There are reports of them drowning by the score in maple sap buckets, and Collembola have been found eating pollen and the fibers and spores of certain fungi. But what could they be eating on clean-looking snow? Admittedly, pollen grains are pretty hard to spot with the naked eye, but nothing I know of releases pollen in winter. You can find certain kinds of fungi or algae growing on old snow, but I have seen snowfleas gather on freshly fallen powder.

I walked along the footpath leading into a nearby patch of woods and noticed that they were most abundant under hemlocks (*Tsuga canadensis*). Then I noticed that the wet bark of the hemlock trunks, just above the melting snowline, was packed with tiny gray specks, all apparently munching on water-activated green algae flecks—just as they were on my back steps. The wetter portions of fallen hemlock branches that lay in the snow were generally covered with snowfleas; branches of other kinds of trees were not.

It so happened that this particular snowflea emergence began just when my biology students were casting about for research project ideas. By the week's end a pair of students, Catherine and Missy, were busily probing the depths of local snowflea ecology. They chose a nearby patch of woods that sported a convenient hiking trail. Their research program was simple; leave fresh footprints in the snowy trail and record the density of springtails that accumulate in the footprints in different forest habitats. For several sunny mornings in a row, Catherine and Missy stomped out a line of prints and returned on the half hour to count the snowfleas.

It was not the most rigorous project, but it yielded some interesting results. Their footprints quickly filled with Collembola everywhere they went, but the densest accumulations were always among hemlock stands. Mixed conifer stands came in a close second, and birch groves yielded the fewest. The students also recorded slightly lower air temperatures in the shady hemlock and mixed conifer stands. Did these results reflect some inherent patchiness in Collembola populations, or food preferences, or temperature preferences, or what? As is often the case with field research, this brief glimpse into the private lives of snowfleas brought up more questions than it answered.

After a year of observation and reading, I had yet to answer the central question of what the darned things are doing out on the snow in the first place. They certainly did not seem to be eating anything there, because they always seemed to be moving when they were on snow. They were not mating on beds of snow, either, or at least they never let me see them doing it. They seemed to be crossing the snow, apparently after crawling up tree trunks from the forest floor.

One day I chanced upon a twenty-year-old article by Walter H. Lyford entitled "Overland Migration of Collembola (*Hypogastrura nivicola* Fitch) Colonies." At last, I thought, The Answer. I was mistaken. Lyford was writing about migrations on the forest floor itself. In autumn, no less, as well as in spring. It seems that the readily noticed frolics on snow are just loose aggregations, with individuals generally keeping several millimeters of space between themselves and their neighbors. During the autumn and spring, springtails crowd together by the thousands into colonies that travel as discrete units through New England woods.

Lyford followed several colonies through Harvard Forest in September 1971. Each colony was about a foot wide and usually consisted of at least five hundred thousand individuals. (Another author reported colonies three times as wide as this, made up of several million crawling and jumping insects.) Each colony moved in a straight line atop and within the duff of the forest floor, barring the occasional stump and boulder, and traveled as a unit for up to five days at a stretch. Any given swarm typically covered between three and five meters per day. Being a professional scientist, Lyford tried a few experiments. He covered one colony with a black plastic sheet, casting it in shadow. The mob stopped in its tracks and refused to budge until sunlight hit it again. He blocked the paths of colonies with metal barriers sunk through the litter and into the mineral soil below. After one day of immobility, the stymied animals inexplicably shed their skins. Maybe that is what was going on with my friend's apparent snowflea invasion. I wish I had thought to ask her if the things were molting all over her doorstep.

Again, more mysteries. In spite of his detailed analysis of springtail migration, Lyford mentioned no motive for snowfleas to walk across snow. Given that they might get together in spring and fall for an occasional molting orgy, why would little black insects leave the relative warmth and safety of a forest floor to scamper across icy white expanses?

Finally, after months of phone calls, I located someone who had a firm, if rather disappointing, answer to the mystery. Dr.

Kenneth Christiansen, a professor at Grinnell College, happened
to be in his office when I called for the fifth time in as many
months. He generously agreed to let me interrogate him on the
subject of springtails, his research specialty over the last several
decades.

So, what are snowfleas doing out there on the snow in the
middle of winter? As Christiansen put it: "Nothing."

"They breed in the leaf litter under the winter snow," he explained matter-of-factly. "By midwinter their populations get so
big that they just sort of boil out onto the surface of the snow. The
ones you see outside like that are just the unlucky ones that got
squeezed out of the crowded upper soil layers. They aren't doing
anything out there; there's nothing for them to eat on the snow,
they aren't mating, and they have no place in particular to go.
They just sort of wander around all day and work their way back
down to the soil around the bases of tree trunks for the night.
Basically, they're pretty unhappy little characters."

Christiansen was full of fascinating tales of springtail behavior
and ecology that almost made up for the let-down of his snowflea
explanation. Collembola live almost anywhere, often in huge
numbers; some say that they may be the most numerous land
animals on Earth, after nematodes. There are parts of Antarctica
where the only vegetation is lichens, the only animals are springtails, and the soil is made almost entirely of springtail droppings.
Although springtails are harmless to humans, they sometimes
become aggressive with each other over disputed territories in
the soil. One of their standard methods of attack is to bludgeon
each other with their floppy, jointed antennae, using them like
clubs with which to bop each other over the head. It does not
seem to hurt them much, but it serves the purpose of expressing
displeasure: "Sometimes you find a pair of them chasing each
other around and around in circles, beating each other's behinds

with their antennae as they run." Oddly, this aggressive side of springtails seems to vanish when they get crowded, as in the winter population booms.

As I thanked Dr. Christiansen and hung up the phone, I felt a mix of relief—in finally getting an answer to the snowflea mystery from a world expert,—and disappointment, as if the pot of gold at the end of the rainbow was full of sand. Now I am going to have to feel sorry for the little creatures when I see them in their teeming masses, wandering like homeless refugees across the frozen wastes of winter.

Maple Sap

If you try hard enough, you can make sugar and syrup from the
sap of many kinds of tree. Red maple *(Acer rubrum)* yields mildly
sweet sap, but it is not very sugar-rich (normally only about 1 to
2 percent sugar), so it is not normally worth the effort of collecting
it and boiling it down. The same goes for most other North American maples. You can also collect a serviceable liquid from black
birch trees *(Betula
lenta)* in spring;
birch beer

is traditionally made from fermented black birch sap. An old friend once showed me how to scrape the pulpy wet inner bark and freshly exposed wood of yellow birch *(Betula alleghaniensis)* with a pocket knife and to lick the syrupy sweet mush from the flat of the blade.

But only the sugar maple *(Acer saccharum),* native to the northeastern United States, makes a sap that is concentrated enough to attract humans on a large scale. Even so, you still have to boil off thirty or forty gallons of water to get a gallon of sugar maple syrup.

The gathering of the raw liquid can be done at most any time between January and bud burst, depending on the weather. If you would like to try it on some of your own trees, all you need to do is to drill some three-inch-deep holes on the sunny sides of the trunks, about three feet above ground level. Make sure that the holes tilt slightly, so the sap flows out easily. Tap a spout, or *spile,* gently into each hole, hang a pail from a notch or hook on the spile, and wait for the flow to begin.

Once you have about forty gallons of raw sap, you are ready to spend the whole day beside your stove, watching the stuff boil down into a single jug's worth. You can use your own judgement in deciding when it is finished (using the licked spoon test), or you can take the temperature of the syrup. The proper concentration of sugar in the liquid raises the boiling temperature to seven degrees above that of pure water.

Maple sap is sweet when you boil it down because it contains sucrose sugar molecules made by the leaves. Sugar is the fuel source that plants use for growth and repair, much as we use glucose sugar in our blood for energy. You might even think of maple sap as a form of colorless plant blood, although it lacks the oxygen-bearing red cells that pigment our blood. It may seem odd at first to think of maple trees consuming their own sap, but

that is what they do. Plant cells burn sugar to generate the energy
they need for producing stems, leaves, flowers, fruits, and so on.
One major difference between plant and animal cells is in the
source of their nutritional fuels; plants make their own sugars
from air, water, and sunlight, but animals must steal theirs from
other organisms in the form of edible body parts.

I like to think of my role in the recycling of molecules when I
eat fresh maple syrup at the "sugar bush" operated by Paul Smith's
College, here in the northern Adirondacks. Hans Michielen opens
the spigot on the big wood-fired boiler and pours hot brown
heaven all over my plate. I take a deep breath of steamy sap-
sweetened air, bringing plant-generated oxygen to my cells. They
will use that oxygen to burn up the maple sugar that slides down
my throat in a mouthful of pancake. The maple trees rustling
around us made these sugar molecules from carbon dioxide, and
when my cells are through with the sugars, I exhale them back into
the air as carbon dioxide again. Some of my breath may eventually
end up back in the very trees that yielded the sap. The sunlight
energy that last year's maple leaves captured to hitch the original
molecules together remains with me to run my muscles and nerves.

Although most plants make sucrose through photosynthesis
for their own consumption, only a few make it in large enough
amounts to interest us much. The sparkling sucrose grains that
we call "table sugar" are crystallized from the sap-rich stalks of
sugar cane *(Saccharum officinarum)*, a large tropical grass native to
Southeast Asia, or from crushed sugar beets *(Beta vulgaris)*. Corn
syrup (from *Zea mays*), the stuff that you often get when you do
not feel like paying for pure maple pancake syrup, comes from
the sap of corn stalks.

The green cells in maple leaves gather sunlight during the
growing season and harness its energy to turn airborne carbon
dioxide molecules into sugar molecules. The sugar leaks out of

those cells and into tiny vessels in the leaf that run together to form small veins that meet in one big central vein at the base of the leaf. This central vein is full of microscopic, fluid-filled tubes, all clustered tightly together like the individual wires in a telephone cable. Some of the tubes in the leaf veins, *xylem* (pronounced "ZI-lum"), bring water and soil nutrients in to the leaf tissues from the roots. Xylem tubes are shaped like long, thin drinking straws, and they are not just restricted to the leaves; they also make up the bulk of a tree's wood. Other vein tubes, *phloem* (pronounced "FLOW-m"), drain newly made sugary sap out of the leaf and into the bark of the woody twig. From there it spreads to the trunk and roots. Phloem tubes outside the leaf make up the pale, moist inner bark that rests flush against the smooth wood of the tree. They are also somewhat straw-shaped, but they tend to have pointier ends than do xylem tubes, and they have little dividers in them with tiny holes that let sap pass from section to section.

It is easy to forget that fluids flow continuously through the twigs, leaves, and limbs of the Northern Forest during the warmer months of the year, but sometimes we are reminded of it accidentally. That happened to me on a warm, buggy day in May. I arrived on the shore of the Little River in the western Adirondacks, fly rod in hand, ready to practice my casting. On the first cast, I hooked my fly in an overhanging yellow birch branch behind me. (Naturally, the black flies immediately took full advantage of my preoccupation with retrieving the fly intact.) I could see the tiny thing stuck to a twig among the bursting leaf buds, so I reached up and pulled the whole branch down towards me. The branch was not as sturdy as I had thought, though, and it kinked and broke close to the trunk as I yanked on it. This yielded the fly, but I felt bad about the damage, knowing that harmful fungi could infect such a wound very quickly.

Turning my back on the tree to resume my pitiful attempts at casting, I forgot about the branch until a dripping sound behind me caught my attention. Clear, watery sap was gushing out of the wound; if birch trees had red sap, it would have been positively gruesome. Although I felt even sorrier for the tree after that, I also appreciated the lesson in tree physiology. Powerful fluid pressures strain in those silent woody limbs, as tree roots press life-giving water and sweetness up and out to their newly sprouting buds.

With so many people enamored of maple syrup and sugar and willing to pay top dollar for it, you might think that every aspect of maple sap production has been thoroughly investigated and explained by scientists by now. In fact, much of the process of sap flow in maples (and in most other plants, for that matter) remains a mystery. This, of course, is not meant to imply that people do not know anything about maple sap. Its chemical composition has been thoroughly analysed and shown to consist mainly of water and sweet sucrose with small amounts of other sugars, vitamins, minerals, and plant growth hormones, a dash of microbe-killing formaldehyde, and even the occasional drifting plant virus mixed in.

Because of its antimicrobial chemical defenses, maple sap is virtually sterile as long as it remains inside the vessels of the host tree. But as soon as you collect it in a nonsterile manner (including most tapping practices), microorganisms can and do invade the liquid from the surrounding air, bark, sampling equipment, and your own body. Microbial activity in commercial maple syrup is generally harmless to humans, but it does noticeably darken the syrup. Tappers who want to keep most of their syrup in the form of premium "light amber grade" (as opposed to "medium" or "dark amber grade") need to keep their tap-holes and equipment scrupulously clean and to store the sap at low temperatures to discourage microbial growth.

The tissues that carry the precious liquid throughout the tree have been located, named, and described in detail. Phloem tubes making up the innermost layers of bark carry most of it, although in late winter and early spring, some of the woody xylem vessels carry it as well. But the precise mechanism by which sucrose gets from one part of the tree to another has been frustratingly elusive.

Many experiments have been carried out in the quest for the mysterious sap transport mechanism. Researchers are pretty sure that it is not just a slow diffusion from roots to buds. In spring, the sap seems to rise through the phloem tubes under pressure as water and minerals would in the xylem, because it flows freely out of wounds in the trunk and branches. Estimates of sap velocity rising in phloem tubes range from about six inches to about ten feet per hour; most are closer to one or two feet per hour.

The sweet sap found in the outermost woody xylem tubes of birch tree trunks in spring rises under pressure provided by molecular pumps and chemical gradients in the roots, but this may not be true for sugar maples. Tree tappers know that maple sap flows fastest on warm days in late winter and stops moving when the sun goes down each evening. This observation has led some to suggest that the slight thermal expansion of a maple's trunk during the warm daylight hours may open xylem tubes more widely than in the cold of night. In this case, fluids could flow through the larger tubes during the day but would be squeezed off at night. But this still does not fully explain what makes sugars move up through maple xylem, much less through the maple phloem.

More confusing still, the sap in phloem does not always flow upwards, as fluids in woody xylem tubes do. When summer leaves are at work, they send sap down the phloem tubes into the trunk and roots. And sometimes it seems to flow in two directions at once. In one experiment conducted in 1930, an investigator cut

two girdle rings into the bark of an apple tree branch, thus completely removing the phloem-containing bark from each strip and
leaving the bare wood exposed. The developing apples that were
thus isolated both from the roots and from the leaves by the upper
and lower cuts withered, as expected. But the ones attached above
and below the cuts continued to grow unaffected. The lower ones
were supplied with sugar by phloem fluids from the roots, and
the upper ones by phloem fluids from the leaves.

All in all, phloem is pretty strange stuff. For one thing, it is
alive when it first forms. A phloem tube lives for several months,
then collapses. Dead, flattened phloem builds up in the older,
outer layers of bark, but it does not conduct sap very well, if at
all. Most sap transport in phloem takes place in a narrow band of
young, living tubes that develop each spring in the innermost
millimeter or so of bark.

Fortunately, our inability to understand phloem completely yet
does not keep maple sap from rising anyway. In late winter and
early spring, the buds at the ends of the twigs get ready to open
and to grow into a new set of leaves, and as a result their bur-
geoning cells need a great deal of sugar-based energy. Because
there are no leaves out at that time of year to manufacture sugar,
the only place to get sucrose for growth energy is from storage
areas in the roots. So the tree pumps sweet sap up into the branches
to feed the swelling buds. If you drill a hole into the tree's trunk
at that time, you will intercept some of the sap on its way up, and
it will drain into your sap bucket. Once the buds burst in spring,
the main sap run is over. But sap still flows well into late spring,
until the leaves are about half grown and can produce enough
sugar to feed themselves. After that, the flow of sap in the phloem
tubes reverses as the leaves become sugar exporters to the root
storage zones.

It is easy to tell when the sap is rising in the sugar maples.
After a frosty night in late winter, when the sun shines brightly

and warms the trees, sap flows freely through the outer layers of the trunks. Keep an eye on the gray-brown bark and on the tips of twigs that were broken during the winter. As soon as you see dark wet stains running from wounds in the bark, or creamy white icicles hanging off the ends of the broken twigs, you know that the sap has begun to flow.

Go up to one of those hanging "sapsicles" and gently pull it down to your mouth. Nibble off the very tip, and you will get a burst of maple flavor on your tongue. The water in the sap froze in the cold night air and left the sugar behind to settle into the tip of the icicle. That saves you from having to boil the stuff down to get a taste. The sugar in the drop at the very tip of the sapsicle is often so concentrated that it lowers the freezing temperature of the liquid, and you get sweet mush on the end of solid ice. Chickadees also take advantage of the moisture and sugar treats in sapsicles on occasion, hovering like hummingbirds to catch the droplets from the slushy tips or snatching away the whole sapsicle to eat on a secure perch.

If you want more sugar than that, you will have to go through a great deal of work. Sap is so dilute that it tastes like water (maple sap fresh from the tree has only about 2 or 3 percent sucrose in it). As a result, you have to boil a lot of water out of it to concentrate it to the minimum 66 percent sucrose content of maple syrup. Getting it all the way down to sugar is even more work. No wonder real maple syrup and candies are so expensive.

But, of course, the taste is worth the trouble. Native Americans in sugar maple country used to collect and boil the sap down in birch bark containers and to mix the sugar with wild rice or pour the syrup onto snow to make maple snow cones. They introduced it to white settlers, and thus began a world-wide affection for The Real Stuff.

Humans and chickadees are not the only animals to have figured out how good maple sugar is. The little gray moths that awaken before the snow is gone drown by the score in open sap pails. Occasionally, a flying squirrel finds a watery grave in a sap bucket, one more reason to check your buckets often and to keep them covered. Yellow-bellied sapsuckers (*Sphyrapicus varius*) and their relatives drill, maintain, and guard "sap well" holes in the phloem of many species of North American tree, and they do it even in summer. Their tongues are tipped with delicate, fibrous projections that help them to sponge up the nutritious liquid from the depths of their woody wells. Other birds sometimes pilfer sap from their wells, when they can get away with it.

And there is one very common nonhuman mammal that not only loves maple sugar but also knows how to get it out of the tree. Some of the oozing bark wounds that you find on maple trunks and branches may have been made on purpose. Red squirrels discovered the sugar maple's little secret long before we did, and they go for the goods when the sap is rising. Watch them at the right time of year and you might spot one nibbling into the bark or branch tips. Once the sap leaks out of the wounds, a few hours of evaporation in the sun's warmth concentrates the stuff enough to give the squirrels something worth licking up.

Bernd Heinrich, a noted zoologist and author at the University of Vermont, made some of the first detailed observations of maple tapping by red squirrels in southern Maine in 1990. He watched several squirrels as they scampered around in a grove of young sugar maples and nibbled small grooves through the bark and outer layers of wood on trunks and branches. Sap oozed from the wounds even in January, forming long wet streaks on the rough bark. Returning each day, after the sun had done its work, the squirrels lapped up concentrated sap from the wet streaks. Heinrich measured the sugar content of the "candied" sap and found sugar concentrations richer than the 55 percent limit of his measuring device. The squirrels had made full-fledged syrup.

The more profuse the branching and foliage are, the more sugar the tree is likely to yield. It would be interesting to learn whether red squirrels prefer to use the bushiest trees. Some red squirrels do seem to follow regular maple traplines during sugar season, scampering from tree to tree in search of the most productive ones. Some folks speculate that the Native Americans who turned the rest of the world on to maple sugar originally learned about tree sweets by watching the squirrels. Or maybe it was the sap-suckers or the chickadees, or just some inquisitive kid who liked to eat icicles. Wherever the idea originally came from, I am glad that we all get to share in the benefits.

Postscript

It has been five years since I began work on these essays at the suggestion of my friend Maurice Kenny. The long time span has allowed the ideas and the wording to evolve and mature naturally through many rewrites. It has also allowed me to reflect on the place of such a book in the general scheme of things.

In discussing this project with others over the years, I have encountered some interesting questions that shed light on why this kind of writing is worth pursuing (certainly, it is not for the money). Among them is: Why must you probe into the scientific details behind the natural beauty of the northern lights? Doesn't it take away from the magic and mystery of the experience? My emphatic response is that accurate and relevant knowledge can only enhance one's appreciation of the natural world. I liken ignorance of the natural world to watching a play with all of its dialogues given in a language with which you are unfamiliar, or to listening to a symphony without any knowledge of the composer's life, the instruments, the musical structure, or the theme of the piece. Certainly such an experience can be pleasurable, but not nearly as rich as that gained through even the most rudimentary background knowledge.

An equally important but more disturbing question is: Who cares about ground bees? With so much of the natural world under assault by humankind, isn't it more important to focus on halting the destruction of endangered habitats and species than on such mundane subjects? I fully agree that all of us need to do what we can to minimize our harmful impacts on the natural world. But rather than add to the atmosphere of gloom so common in environmental media today, I seek to provide a note of hope and some carefully researched information to help guide and support the efforts of otherwise well-intentioned but often poorly informed activists. How can we be fully effective stewards of the natural world if we don't understand what's going on in our own backyards? Now is not the time to be preoccupied with despair over what has been lost; far more remains of the natural world than many of us realize, especially on the smaller size scales. My simple backyard encounters with ground bees have turned a college lawn into an informal wildlife sanctuary, and now that the decline of domestic bees owing to mite infestations is big news in North American agriculture, my students and I can appreciate how it enhances the significance of wild native pollinators.

Seeking awareness of the natural world is a noble and rewarding endeavor, but today it is much more than that. I strongly believe that it is a personal responsibility. *Homo sapiens* not only covers the land habitats of the earth; we are now linked electronically to each other. Televisions, radios, telephones, faxes, and computers permit ideas, attitudes, and plans to alter human awareness and actions on massive scales and at tremendous speeds. In my view, human thought is now one of the most powerful environmental forces on this planet; just consider the global environmental and economic consequences of the simple preference for driving personal automobiles rather than sharing public transportation. What you carry in your head affects how you act, and with billions of us thinking and acting (and doing so more and more frequently in unison), it is vital that the information upon which we base our behavior be as accurate and complete as possible and that our attitudes be as humane and as responsible as possible.

In this new global context, learning itself can be a form of environmental activism. I urge you to count yourself among those who know what lies underground, underwater, and underskin; who are familiar enough with the diversity of their nonhuman neighbors to speak of them by name; who know what is lost when a "vacant" lot is paved; who fall ever deeper in love with this magical world as they come to know it ever more intimately; and who are willing to defend it, gently but firmly, from those who are unwilling or unable to see.

Glossary
References
Index

Glossary

Ameiurus nebulosus: lit., "dark joined tail" (?). Greek *ama*, "together," Greek *eiro*, "join, connect," and Greek *oura*, "tail"; Latin *nebulosus*, "dark, clouded."

Ant, Black Carpenter: *Camponotus pennsylvanicus*, lit., "Pennsylvanian bending-back." Greek *campo*, "bending" or "flexible" or "caterpillar," and Greek *noto*, "back."

Armillaria bulbosa: lit., "bracelet of bulbs." Latin *armilla*, "bracelet"; Latin *bulbosa*, "full of bulbs."

Armillaria clavescens: lit., "slightly clubbed bracelet." Latin *armilla*, "bracelet"; Latin *clava*, "club," and Latin *-escens*, "slightly."

Aspen, Quaking: *Populus tremuloides*, lit., "trembling multitude." Latin *populus*, "people, a great number"; Latin *tremulus*, "trembling."

Baragwanathia: probably named after a place or a person?

Bass, Largemouth: *Micropterus salmoides*, lit., "salmonlike small fin." Greek *mikros*, "small," and Greek *pteron*, "fin" or "wing"; Latin *salmo*, "salmon," and New Latin *-oides*, "like."

Bayberry: *Myrica pennsylvanica*, lit., "Pennsylvanian tamarisk." Greek *myrike*, "tamarisk."

Bear, Black: *Ursus americanus*, lit., "American bear." Latin *ursus*, "bear."

Bear, Grizzly: *Ursus horribilis*, lit., "terrible bear." Latin *ursus*, "bear"; Latin *horrib*, "terrible" or "fearful."

219

Bear, Polar: *Ursus maritimus,* lit., "sea bear." Latin *ursus,* "bear"; Latin *maritim,* "of the sea."

Beaver: *Castor canadensis,* lit., "Canadian beaver." Greek *kastor,* "beaver."

Beaver, European: *Castor fiber,* lit., "beaver beaver." Greek *kastor,* "beaver"; Latin *fiber,* "beaver."

Bee, Cuckoo: *Nomada,* lit., "wanderer." Greek *nomad,* "roving, spreading."

Bee, Honey: *Apis mellifera,* lit., "honey-bearing bee." Latin *apis,* "bee"; Latin *mellis,* "honey," and Latin *fero,* "to bear."

Beech: *Fagus grandifolia,* lit., "large-leafed beech." Latin *fagus,* "beech"; Latin *grandi,* "large," and Latin *folium,* "leaf."

Beet: *Beta vulgaris,* lit., "common beet." Latin *beta,* "beet"; Latin *vulgaris,* "common."

Beetle (Order): Coleoptera, lit., "sheath wing." Greek *koleos,* "sheath," and Greek *pteron,* "wing."

Beetle, Fire: *Pyrophorus,* lit., "fire carrier." Greek *pyr,* "fire," and Greek *phoros,* "bearing."

Bigtooth aspen: *Populus grandidentata,* lit., "big-toothed multitude." Latin *populus,* "people, a great number"; Latin *grandis,* "big" or "large," and Latin *dentatus,* "toothed."

Birch: *Betula,* lit., "birch." Latin *betula,* "birch."

Birch, Black: *Betula lenta,* lit., "sticky birch." Latin *betula,* "birch"; Latin *lentus,* "sticky, slow, tough, pliant, tenacious."

Birch, White: *Betula papyrifera,* lit., "paper-bearing birch." Latin *betula,* "birch"; Greek *papyros,* "paper-reed," and Latin *fero,* "to bear."

Birch, Yellow: *Betula alleghaniensis,* lit., "Alleghenian birch." Latin *betula,* "birch."

Blackbird, Redwing: *Agelaius phoeniceus,* lit., "crimson gregarious." Greek *agelaios,* "gregarious"; Greek *phoinix,* "crimson."

Blueberry: *Vaccinium,* lit., "cowberry." Latin *vaccin,* "of a cow."

Bluet: *Houstonia caerula,* lit., "Houston's blue." Latin *caeruleus,* "dark blue."

Boletus: lit., "mushroom." Greek *Bolites,* a kind of mushroom.

Borer, Maple: *Glycobius speciosus,* lit., "beautiful sweet life" (?). Greek *glykos,* "sweet," and perhaps Greek *bios,* "life"; Latin *specios,* "showy, beautiful."

Bug (Order): Hemiptera, lit., "half wing." Greek *hemi*, "half," and Greek *pteron*, "wing."

Bug, Giant Water: *Lethocerus*, lit., "deadly horn." Latin *lethalis*, "deadly," and Greek *keras*, "horn."

Cabbage, Skunk: *Symplocarpus foetidus*, lit., "stinky connected fruit." Greek *symploc*, "connected," and Greek *carpus*, "fruit"; Latin *foeten*, "evil-smelling."

Calypso bulbosa: lit., "bulbous nymph." Greek *calypso*, "nymph"; Latin *bulb*, "bulb," and Latin *osa*, "full of."

Cane, Sugar: *Saccharum officinarum*, lit., "sugar storeroom." Greek *sakchar*, "sugar"; Latin *officina*, "storeroom."

Caribou: *Rangifer tarandus*, lit., "wild reindeer reindeer." Old Swedish *ren*, "reindeer," and Latin *fera*, "wild beast," converted to New Latin *rangifer*, "reindeer"; Greek *tarandos*, "reindeer."

Castoroides ohioensis: lit., "beaverlike animal from Ohio." Greek *kastor*, "beaver," New Latin *-oides*, "like."

Cattail: *Typha*, lit., "cattail." Greek *typhe*, "cattail."

Cherry: *Prunus*, lit., "plum." Latin *prunum*, "plum."

Cherry, Black: *Prunus serotina*, lit., "late-ripening plum." Latin *prunum*, "plum"; Latin *serotinus*, "late-ripening, backwards."

Chestnut: *Castanea dentata*, lit., "toothed chestnut." Greek *kastanos*, "chestnut"; Latin *dentatus*, "toothed."

Chickadee: *Parus atricapillus*, lit., "black-haired titmouse." Latin *parus*, "titmouse"; Latin *atri*, "black," and Latin *capill*, "hair."

Chipmunk: *Tamias striatus*, lit., "striped gatherer." Greek *tamia*, "distributor" or "gatherer"; Latin *stria*, "furrow" or "channel."

Colletes inaequalis: lit., "unequal glue-dweller." Greek *coll*, "glue," and Greek *-etes*, "dweller"; Latin *inaequali*, "unequal, uneven."

Comfrey: *Symphytum officinale*, lit., "official joined plant" (?). Greek *symphy*, "to glue together," and perhaps Greek *phyton*, "plant"; Latin *officinalis*, "of an office."

Copperhead: *Agkistrodon mokasen*, lit., "moccasin fang" (?). Latin *ag- (ad)*, "toward," perhaps Greek *kiste*, "chest, box," and Greek *odontos*, "tooth"; Native American *moccasin*, "shoe."

Corn: *Zea mays*, lit., "corn grain." Greek *zea*, "grain"; Native American *maize*, "corn."

Cougar: *Felis concolor*, lit., "cat with color." Latin *felis*, "cat"; Latin *con-*, "with," and Latin *color*, "color."

Coyote: *Canis latrans*, lit., "barking dog." Latin *canis*, "dog"; Latin *latrans*, "barker."

Cranberry: *Vaccinium*, lit., "whortleberry." Latin *vaccinium*, "whortle-berry."

Cuckoo, European: *Cuculus canorus*, lit., "melodious cuckoo." Latin *cuculus*, "cuckoo"; Latin *canor*, "song, melody."

Dandelion: *Taraxacum officinale*, lit., "official dandelion." Persian *tarashquin*, "dandelion"; Latin *officinalis*, "of an office."

Deer, Whitetailed: *Odocoileus virginianus*, lit., "Virginian hollow-tooth." Greek *odous*, "tooth," and Greek *koilos*, "hollow."

Elm, American: *Ulmus americana*, lit., "American elm." Latin *ulmus*, "elm."

Eucalyptus: lit., "well covered" or "good cover." Greek *eu*, "good," and Greek *kalyptos*, "covered."

Exyria rolandiana: lit., "Roland's consumer" (?). Latin *exedo*, "to consume."

Fern, Royal: *Osmunda regalis*, lit., "royal god." Saxon *Osmunder*, a god; Latin *regalis*, "royal."

Fir, Balsam: *Abies balsamea*, lit., "balsam fir tree." Latin *abies*, "fir tree"; Latin *balsamum*, "balsam."

Fish, Lantern: *Photoblepharon*, lit., "light eyelid." Greek *photos*, "light," and Greek *blepharon*, "eyelid."

Flavoparmelia: lit., "yellow shield." Latin *flav*, "yellow," and Latin *parm*, "shield."

Flicker, Northern: *Colaptes auratus*, lit., "golden hammerer." Greek *kolapter*, "chisel" or "peck" or "hammer"; Latin *auratus*, "golden."

Fly (Order): *Diptera*, lit., "two wing." Greek *di-*, "two," and Greek *pteron*, "wing."

Flytrap, Venus's: *Dionaea muscipula*, lit., "divine fly-flea." Greek *dion*, "divine"; Latin *musci*, "fly," and Latin *pulex*, "flea."

Frog, Wood: *Rana sylvatica*, lit., "woodland frog." Latin *rana*, "frog"; Latin *sylva*, "a woodland."

Ginger, Wild: *Asarum canadense,* lit., "Canadian stemless shrub." Greek *asaron,* "a low, stemless shrub."

Goldenrod: *Solidago,* lit., "golden rod." Middle Latin *solidago,* "golden-rod."

Goldfinch: *Carduelis tristis,* lit., "melancholy of the thistle." Latin *carduus,* "thistle," and Latin *-elis,* "of"; Latin *tristis,* "melancholy, sad."

Grackle: *Quiscalus quiscula,* lit., "quail quail." Middle Latin *quiscula,* "quail."

Grouse, Ruffed: *Bonasa umbellus,* lit., "umbrella bird." Latin *bonasa,* a type of bird; Latin *umbella,* "umbrella."

Hare, Snowshoe: *Lepus americanus,* lit., "American hare." Latin *lepus,* "hare."

Hemlock: *Tsuga canadensis,* lit., "Canadian hemlock." Japanese *tsuga,* "larch" or "hemlock."

Hepatica americanum: lit., "American liver." Greek *hepatikos,* "liver."

Heron, Great Blue: *Ardea herodias,* lit., "heron heron." Latin *ardea,* "heron"; Greek *herodios,* "heron."

Hypogastrura nivicola: lit., "snow-dwelling under-belly." Greek *hypo,* "under" or "beneath," and Greek *gaster,* "stomach" or "belly"; Latin *nivis,* "snow," and Latin *colo,* "to inhabit."

Indian Pipe: *Monotropa uniflora,* lit., "singly bent single flower." Greek *mono,* "one" or "single," and Greek *tropos,* "a turn or change"; Latin *uni,* "one," and Latin *flora,* "flower."

Ivy, Poison: *Rhus radicans,* lit., "rooting sumac." Latin *rhus,* "sumac"; Latin *radicans,* "taking root."

Jack-in-the-Pulpit: *Arisaema triphyllum,* lit., "three-leafed blood arum." Latin *aris,* "arum plant" and Greek *haema,* "blood"; Latin *tri,* "three," and Latin *phyllum,* "leaf."

Jay, Blue: *Cyanocitta cristata,* lit., "blue crested chattering bird." Greek *kitta,* "chattering bird," and Greek *kyaneos,* "dark blue"; New Latin *cristata,* "crested."

Kingfisher, Belted: *Ceryle alcyon,* lit., "kingfisher kingfisher." Greek *kerylos,* "kingfisher"; Greek *alkyon,* "kingfisher."

Kinglet, Ruby-Crowned: *Regulus calendula,* lit., "little king of the month." Latin *regulus,* "prince" or "little king"; Latin *Kalendae,* "first day of the month."

Labrador Tea: *Ledum groenlandicum,* lit., "Greenland shrub." Greek *ledon,* "mastic, shrub."

Lady's Slipper, Pink: *Cypripedium acaule,* lit., "goddess slipper." Greek *Kypris,* a goddess of love, and Latin *pedis,* "foot"; Greek *a-,* "without" and Latin *caulis,* "stalk" or "stem."

Larch: *Larix laricina,* lit., "larch of the larch." Latin *larix,* "larch"; Latin *laricinus,* "pertaining to the larch."

Laurel, Bog Sheep: *Kalmia angustifolia,* lit., "Kalm's narrow-leaf." Latin *angustus,* "narrow, small," and Latin *folium,* "leaf."

Lepidodendron: lit., "scale tree." Greek *lepido,* "scale," and Greek *dendron,* "tree."

Lichen, Reindeer: *Cladina,* lit., "branches." Greek *klados,* "branch" or "sprout."

Lily, Trout: *Erythronium americanum,* lit., "American orchid." Greek *erythronion,* an orchid.

Lycopodium: lit., "wolf foot." Greek *lykos,* "wolf," and Greek *podos,* "foot."

Mammoth: *Elephas primigenius,* lit., "original elephant." Greek *elephas,* "elephant"; Latin *primus,* "first," and Greek *genos,* "a kind."

Maple, Red: *Acer rubrum,* lit., "red maple." Latin *acer,* "maple" or "sharp"; Latin *ruber,* "red."

Maple, Sugar: *Acer saccharum,* lit., "sugar maple." Latin *acer,* "maple" or "sharp"; Greek *sakchar,* "sugar."

Marigold, Marsh (or Cowslip): *Caltha palustris,* lit., "marsh marigold." Latin *caltha,* "marigold"; Latin *palus,* "marsh."

Mastodon: *Mastodon americanus,* lit., "American breast-tooth." Greek *mastos,* "breast," and Greek *odontis,* "tooth."

Mink: *Mustela vison,* lit., "weasel weasel." Latin *mustela,* "weasel"; Icelandic or Swedish *vison,* a kind of weasel.

Moose: *Alces alces,* lit., "elk elk." Greek *alke,* "elk."

Mosquito: *Aedes hendersoni,* lit., "Henderson's disagreeable." Greek *aedes,* "disagreeable."

Mosquito: *Anopheles,* lit., "flying ensnarer." Greek *ano-,* "upward, aloft," and Greek *pheloo,* "to ensnare or deceive."

Mosquito: *Culex,* lit., "gnat." Latin *culex,* "gnat, small fly, midge."

Moss, Haircap: *Polytrichum,* lit., "many hairs." Greek *poly,* "many," and Greek *trichos,* "hair."

Moth, Gypsy: *Lymantria dispar,* lit., "different destroyer." Greek *lymanter,* "destroyer"; Latin *dispar,* "different, unlike."

Musk-Ox: *Ovibos moschatus,* lit., "musky sheep-ox." Latin *ovis,* "sheep," and Latin *bous,* "ox"; Greek *oschatus,* "musky."

Muskrat: *Ondatra zibethica,* lit., "civet muskrat." Native American *ondatra,* "muskrat"; Greek *zibeth,* "civet."

Nightshade, Black: *Solanum nigrum,* lit., "deadly comforter." Latin *solor,* "comfort, soothe"; Latin *niger,* "deadly, black."

Nuthatch, White-Breasted: *Sitta carolinensis,* lit., "Carolinian nuthatch." Greek *sitte,* "nuthatch."

Nutria: *Myacastor,* lit., "mouse beaver." Greek *mys,* "mouse," and Greek *kastor,* "beaver."

Oak: *Quercus,* lit., "oak." Latin *quercus,* "oak."

Onchocerciasis: lit., "hook-tail disease." Greek *oncho,* "hook" or "barb," and Greek *cercus,* "tail."

Otter: *Lutra canadensis,* lit., "Canadian otter." Latin *lutra,* "otter."

Penicillium: lit., "pencil." Latin *penicillum,* "pencil."

Pepper, Red: *Capsicum,* lit., "of the box." Latin *capsa,* "box" or "case," and Latin *-icum,* "belonging to."

Perch, Yellow: *Perca flavescens,* lit., "yellowing perch." Greek *perke,* "perch"; Latin *flavescens,* "becoming yellow."

Photinus carolinus: lit., "Carolinian light." Greek *photeinos,* "light, shining."

Photinus collustrans: lit., "neck-encircling light." Greek *photeinos,* "light, shining"; Latin *collum,* "neck," and Latin *lustrans,* "encircling."

Photinus macdermotti: lit., "Macdermott's light." Greek *photeinos,* "light, shining."

Photinus pyralis: lit., "fire light." Greek *photeinos,* "light, shining"; Greek *pyralis,* an insect that lives on fire, or Greek *pyr,* "fire."

Pine, Bristlecone: *Pinus aristata,* lit., "brittle pine" or "best pine." Latin *pinus,* "pine"; Latin *arista,* "brittle," or Greek *aristos,* "best."

Pine, Loblolly: *Pinus taeda,* lit., "pine pine." Latin *pinus,* "pine"; Latin *taeda,* a kind of pine.

Pine, Longleaf: *Pinus palustris,* lit., "swamp pine." Latin *pinus,* "pine"; Latin *palustris,* "swampy" or "marshy."

Pine, Ponderosa: *Pinus ponderosa,* lit., "ponderous pine." Latin *pinus,* "pine"; Latin *ponderosus,* "ponderous, weighty."

Pine, Red: *Pinus resinosa,* lit., "resinous pine." Latin *pinus,* "pine"; Latin *resina,* "resin."

Pine, White: *Pinus strobus,* lit., "twisted cone pine." Latin *pinus,* "pine"; Greek *strobos,* "twisting, turning, whirling," or Latin *strobilus,* "a pine cone."

Pitcher Plant: *Sarracenia purpurea,* lit., "purple empty sweeper" (?) or "Sarracen's purple." Possibly Greek *saron,* "broom, sweeper," and Greek *kenos,* "empty"; Latin *purpureus,* "purple."

Platanthera dilata: lit., "dilated broad flower." Greek *platos,* "broad, flat," and Greek *anthos,* "flower"; Latin *dilato,* "dilate."

Podura aquaticus: lit., "water-dwelling foot-tail." Greek *podos,* "foot," and Greek *oura,* "tail."

Porcupine: *Erethizon dorsatum,* lit., "aroused with a back." Greek *erethizo,* "arouse, irritate"; Latin *dorsatus,* "with a back."

Potato: *Solanum,* lit., "comforter." Latin *solor,* "comfort, soothe."

Raccoon: *Procyon lotor,* lit., "dog-faced washer." Latin *pro-,* "forward," and Greek *kyon,* "dog"; Latin *lotor,* "washer."

Ragweed: *Ambrosia artemisifolia,* lit., "yarrow-leafed ambrosia." Greek *ambrosia,* the food of the gods; Greek *artemisia,* "yarrow," and Latin *folium,* "leaf."

Raspberry: *Rubus,* lit., "red." Latin *ruber,* "red."

Redpoll: *Carduelis flammea,* lit., "flame of the thistle." Latin *carduus,* "thistle," and Latin *-elis,* "of"; Latin *flammeus,* "flame-colored."

Redwood: *Sequoia sempervirens,* lit., "evergreen sequoia." *Sequoia,* the name of a Cherokee leader; Latin *semper,* "ever," and Latin *virens,* "green."

Salmon, Landlocked: *Salmo salar,* lit., "salmon of salt." Latin *salmo,* "salmon"; Latin *salarius,* "belonging to salt."

Sapsucker, Yellow-Bellied: *Sphyrapicus varius,* lit., "mottled hammer-tip." Greek *sphyra,* "hammer," and Latin *apicus,* "tip" or "point"; Latin *varius,* "mottled, varying, variegated."

Sarcophaga: lit., "flesh eater." Greek *sarco,* "flesh," and Greek *phago,* "eat."

Sassafras: *Sassafras albidum,* lit., "white sassafras." Sassafras may be a modification of Latin *saxifraga,* the saxifrage plant; Latin *albidus,* "white."

Selaginella: lit., "small club moss." Latin *selaginis,* "club moss," and Latin *-ella,* "small."

Sequoia: *Sequoiadendron giganteum,* lit., "gigantic sequoia tree." *Sequoia,* the name of a Cherokee leader, and Greek *dendron,* "tree"; Greek *gigas,* "giant."

Siskin, Pine: *Carduelis pinus,* lit., "pine of the thistle." Latin *carduus,* "thistle," and Latin *-elis,* "of"; Latin *pinus,* "pine."

Snake (Family): Colubridae, lit., "snakes." Latin *colubra,* "snake."

Snake, Common Garter: *Thamnophis sirtalis,* lit., "garterlike shrub snake." Greek *thamnos,* "shrub," and Greek *ophis,* "snake"; New Latin *sirtalis,* "like a garter."

Snake, Ring-necked: *Diadophis punctatis,* lit., "spotted crown snake." Greek *diadem,* "crown," and Greek *ophis,* "snake"; Latin *punctat,* "marked with punctures" or "spotted."

Snake, Timber Rattler: *Crotalus horridus,* lit., "rough rattler." Greek *croto,* "rattle"; Latin *horridus,* "rough, prickly."

Solomon's Seal, False: *Smilacina racemosa,* lit., "like bindweed and clustered" (?). Greek *smilax,* "bindweed, morning glory, yew," and Latin *-ina,* "like"; Latin *racemus,* "cluster, bunch of grapes."

Sorrel, Wood: *Oxalis montana,* lit., "mountain sorrel." Greek *oxalis,* "sorrel"; Latin *montan,* "mountain."

Spartina alterniflora: lit., "alternating cord-grass." Greek *spartine,* "cord, rope"; Latin *alterno,* "alternate, vary," and Latin *floris,* "flower."

Sphagnum cuspidatum: lit., "pointed moss." Greek *sphagnos,* "moss"; Latin *cuspidatus,* "made pointed."

Sphagnum fuscum: lit., "brown moss." Greek *sphagnos,* "moss"; Latin *fuscus,* "brown, dusky."

Sphagnum recurvum: lit., "recurved moss." Greek *sphagnos,* "moss"; Latin *recurvo,* "recurve, bend backwards."

Sphagnum rubellum: lit., "reddish moss." Greek *sphagnos,* "moss"; Latin *rubellus,* "reddish."

Spruce: *Picea*, lit., "spruce." Latin *picea*, "spruce" or "pitch pine."

Spruce, Black: *Picea mariana*, lit., "Maryland spruce." Latin *picea*, "spruce."

Spruce, Red: *Picea rubens*, lit., "red spruce." Latin *picea*, "spruce"; Latin *rube*, "red."

Spruce, Sitka: *Picea sitchensis*, lit., "Sitka spruce." Latin *picea*, "spruce."

Squirrel, Flying: *Glaucomys sabrinus*, lit., "gray mouse nymph." Greek *glaukos*, "gray, silvery," and Greek *mys*, "mouse"; Latin *sabrina*, "a nymph."

Squirrel, Red: *Tamiasciurus hudsonicus*, lit., "Hudson gatherer-squirrel." Greek *tamias*, "distributor or gatherer," and Latin *sciurus*, "squirrel."

Starling: *Sturnus vulgaris*, lit., "common starling." Latin *sturnus*, "starling"; Latin *vulgaris*, "common."

Strawberry, Wild: *Fragaria virginiana*, lit., "Virginian strawberry." Latin *fraga*, "strawberry."

Sumac, Staghorn: *Rhus typhina*, lit., "cloudlike sumac." Latin *rhus*, "sumac"; Greek *typhos*, "cloud, smoke."

Sundew: *Drosera rotundifolia*, lit., "dewy round leaf." Greek *droser*, "dewy"; Latin *rotund*, "round," and Latin *foli*, "leaf."

Sunflower: *Helianthus*, lit., "sun flower." Greek *helios*, "sun," Greek *anthos*, "flower."

Thistle: *Cirsium*, lit., "thistle." Greek *kirsion*, "thistle."

Thrips, Pear: *Taeniothrips inconsequens*, lit., "inconsequential banded woodworm." Greek *tainia*, "band, ribbon," and Greek *thrips*, "woodworm"; Latin *in-*, "not," and Latin *consequent-*, "following."

Tobacco: *Nicotiana tabacum*, lit., "Nicot's tobacco." Jacques Nicot introduced tobacco to France in 1560; New Latin *tabacim*, "tobacco."

Tomato: *Lycopersicon*, lit., "wolf peach." Greek *lykos*, "wolf," and Latin *persica*, "peach."

Trillium: lit., "three tassels (?)." Latin *tri-*, "three," and possibly Latin *licium*, "ends of a weaver's thread."

Trout, Brook: *Salvelinus fontinalis*, lit., "little fountain salmon." New Latin *salvelinus*, "little salmon"; Latin *fontanalis*, "of a fountain or spring."

Trout, Lake: *Salvelinus namaycush*, lit., "little namaycush salmon." New Latin *salvelinus*, "little salmon"; Native American *namaycush*, "lake trout."

Trout, Rainbow: *Onchorhynchus mykiss,* lit., "forked hook-snout"(?). Greek *oncho,* hook, and Greek *rhynchus,* "snout"; Greek *mykiss,* "forked sticks."

Violet: *Viola,* lit., "violet." Latin *viola,* "violet."

White-cedar, Northern: *Thuja occidetalis,* lit. "Western Thuya." Probable modification of Latin *Thuya,* a species of African tree; Latin *occidentalis,* "western."

Willow: *Salix,* lit., "willow." Latin *salix,* "willow."

Woodpecker, Acorn: *Melanerpes formicivorus,* lit., "ant-eating black creeper." Greek *melas,* "black," and Greek *herpo,* "to creep"; Latin *formica,* "ant," and Latin *voro,* "eat."

Woodpecker, Downy: *Dendrocopos pubescens,* lit., "downy tree-toiler." Greek *dendron,* "tree," and Greek *kopos,* "toil, fatigue, unease"; Latin *pubescens,* "fuzzy." Also possibly translatable as "downy tree-striker," using Greek *kolaphos,* "to buffet or strike."

Woodpecker, Green: *Picus viridis,* lit., "green woodpecker." Latin *picus,* "woodpecker"; Latin *viridis,* "green."

Woodpecker, Hairy: *Dendrocopos villosus,* lit., "hairy tree-toiler." Greek *dendron,* "tree," and Greek *kopos,* "toil, fatigue, unease"; Latin *villus,* "hairy." Also possibly translatable as "hairy tree-striker," using Greek *kolaphos,* "to buffet or strike."

Woodpecker, Pileated: *Dryocopos pileatus,* lit., "capped tree-toiler." Greek *dryo,* "tree" or "oak," and Greek *kopos,* "toil, fatigue, unease"; Greek *pileos,* "cap." Also possibly translatable as "capped tree-striker," using Greek *kolaphos,* "to buffet or strike."

Wolf, Timber: *Canis lupus,* lit., "wolf dog." Latin *canis,* "dog"; Latin *lupus,* "wolf."

Wyeomia smithii: lit., "Smith's wyeomia." A person or place?

Xanthoria: lit., "yellow place." Greek *xanthos,* "yellow," and Latin *-oria,* "place of."

References

Introduction

Borror, D. J. 1971. *Dictionary of Word Roots and Combining Forms.* Palo Alto, Calif.: Mayfield.

Ground Bees

Batra, S. W. T. 1990. Bionomics of a Vernal Solitary Bee *Andrena alleghaniensis* Viereck in the Adirondacks of New York, USA. *Journal of the Kansas Entomological Society* 63: 260–266.

Borror, D. J., D. M. Delong, and C. A. Triplehorn. 1976. *An Introduction to the Study of Insects.* New York: Holt, Rinehart, and Winston.

Gebhardt, M., and G. Roehr. 1987. On the Bionomics of the Sand Bees *Andrena clarkella, Andrena cineraria, Andrena fuscipes,* and Their Cuckoo Bees (Hymenoptera, Apoidea). *Drosera* 87: 89–114.

Giblin-Davis, R. M., B. B. Norden, S. W. T. Batra, and G. C. Eickwort. 1990. Commensal Nematodes in the Glands, Genitalia, and Brood Cells of Bees (Apoidea). *Journal of Nematology* 22: 150–161.

Milne, L., and M. Milne. 1980. *The Audubon Society Field Guide to North American Insects and Spiders.* New York: Alfred A. Knopf.

Riddick, E. W. 1992. Nest Distribution of the Solitary Bee *Andrena macra* Mitchell (Hymenoptera, Andrenidae) with Observations on Nest Structure. *Proceedings of the Entomological Society of Washington* 94: 568–575.

Steinmann, E. 1990. On Short-range Orientation of the Solitary Andrenid bee *Andrena vaga* Panzer (Hymenoptera, Apoidae) at the Nest Entrance. *Bulletin de la Société Entomologique Suisse* 63: 77–80.

Stokes, D. W. 1983. *A Guide to Observing Insect Lives*. Stokes Nature Guides. Boston: Little, Brown.

Small Worlds

Borror, D. J., D. M. Delong, and C. A. Triplehorn. 1976. *An Introduction to the Study of Insects*. New York: Holt, Rinehart, and Winston.

Milne, L., and M. Milne. 1980. *The Audubon Society Field Guide to North American Insects and Spiders*. New York: Alfred A. Knopf.

Peterson, R. T., and M. McKenny. 1968. *A Field Guide to Wildflowers*. Boston: Houghton Mifflin.

Woodpeckers

Bent, A. C. 1964. *Life Histories of North American Woodpeckers*. New York: Dover.

Ehrlich, P. R., D. S. Dobkin, and D. Wheye. 1988. *The Birder's Handbook*. New York: Simon and Schuster.

Heinrich, B. 1993. Kinglets' Realm of Cold. *Natural History*, February, 5–9.

Stokes, D. W. 1979. *A Guide to Bird Behavior*. Boston: Little, Brown.

Welty, J. C. 1975. *The Life of Birds*. Philadelphia: W. B. Saunders.

Underground Connections

Allen, M. F. 1991. *The Ecology of Mycorrhizae*. New York: Cambridge Univ. Press.

Barron, G. 1992. Jekyll-Hyde Mushrooms. *Natural History*, March, 47–52.

Brock, T. D. 1966. *Principles of Microbial Ecology*. Englewood Cliffs, N.J.: Prentice-Hall.

Brundrett, M., and B. Kendrick. 1990. The Roots and Mycorrhizae of Herbaceous Woodland Plants, part 1: Quantitative Aspects of Morphology. *New Phytologist* 114: 457–468.

Gould, S. J. 1992. A Humongous Fungus among Us. *Natural History,* July, 10–18.

Knaphus, G., and L. H. Tiffany. 1991. Enchanted Rings. *Insight,* May, 299–303.

Medve, R. J. 1992. Repeat after Me: Indian Pipe Is Not a Saprophyte. *Coltsfoot* 8: 8–10.

Peterson, R. T., and M. McKenny. 1968. *A Field Guide to Wildflowers.* Boston: Houghton Mifflin.

Roden, J. S., and R. W. Pearcy. 1993. Effect of Leaf Flutter on the Light Environment of Poplars. *Oecologia* 93: 201–207.

Thomas, R. J., E. M. Schiele, and D. T. Damberg. 1990. Translocation in *Polytrichum commune,* Bryophyta, part 2: Clonal Integration. *American Journal of Botany* 77: 1569– 1573.

Zimmerman, M. H., and C. L. Brown. 1971. *Trees: Structure and Function.* New York: Springer-Verlag.

Why Do Bugs Bite?

Anderson, R. 1991. How She Gets Your Blood. *Natural History,* July, 43.

Beehler, J. W., J. G. Millar, and M. S. Mulla. 1993. Synergism Between Chemical Attractants and Visual Cues Influencing Oviposition of the Mosquito, *Culex quinquefasciatus* (Diptera: Culicidae). *Journal of Chemical Ecology* 19: 635– 670.

Bodanis, D. 1992. *The Secret Garden.* New York: Simon and Schuster.

Borror, D. J., D. M. Delong, and C. A. Triplehorn. 1976. *An Introduction to the Study of Insects.* New York: Holt, Rinehart and Winston.

Edman, J. D. 1991. Biting the Hand That Feeds You. *Natural History,* July, 8–10.

Klowden, M. J. 1991. Tales of a Mosquito Psychologist. *Natural History,* July, 48–50.

Kukka, C. 1992. My Neighborhood Fights a War with Mosquitos. *Maine Times Spring Guide,* 14–17.

Lindsay, S. 1985. Filariasis Kills Mosquitoes Too. *New Scientist* 106: 24.

Milne, L., and M. Milne. 1980. *The Audubon Society Field Guide to North American Insects and Spiders*. New York: Alfred A. Knopf.

Milner, R. 1991. Can Mosquitoes Transmit AIDS? *Natural History*, July, 54.

Nasci, R. S. 1985. Local Variation in Blood Feeding by *Aedes triseriatus* and *Aedes hendersoni* (Diptera: Culicidae). *Journal of Medical Entomology* 22: 619–623.

Nielsen, L. T. 1991. Mosquitoes Unlimited. *Natural History*, July, 4–6.

Natives

Adams, C. M., and T. C. Hutchinson. 1992. Fine-root Growth and Chemical Composition in Declining Central Ontario Sugar Maple Stands. *Canadian Journal of Forest Research* 22: 1489– 1503.

Bauce, E., and D. C. Allen. 1992. Role of *Armillaria calvescens* and *Glycobius speciosus* in a Sugar Maple Decline. *Canadian Journal of Forest Research* 22: 549–552.

Bhiry, N., and L. Filion. 1996. Mid-Holocene Hemlock Decline in Eastern North America Linked with Phytophagous Insect Activity. *Quaternary Research* 45: 312–320.

Jackson, S. T., and D. R. Whitehead. 1991. Holocene Vegetation Patterns in the Adirondack Mountains, New York, USA. *Ecology* 72: 641–653.

Ketchledge, E. H. 1992. Born-Again Forest. *Natural History*, May, 34–38.

Pollan, M. 1994. Against Nativism. *New York Times Magazine*, May 15, 53–55.

Snakes

Allstetter, B. 1990. Snakes in Drag. *Discover*, June, 20.

Brodie, E. D. 1989. Genetic Correlations Between Morphology and Antipredator Behavior in Natural Popluations of the Garter Snake, *Thamnophis ordinoides*. *Nature* 342: 542–543.

———. 1990. Genetics of the Garter's Getaway. *Natural History*, July, 45–50.

Ford, N. B. 1986. The Role of Pheromone Trails in the Sociobiology of Snakes. In *Chemical Signals in Vertebrates*, edited by D. Duvall, D. Müller-Schwarze, and R. M. Silverstein, 261–278. New York: Plenum.

Garstka, W. R., and D. Crews. 1986. Pheromones and Reproduction in Garter Snakes. In *Chemical Signals in Vertebrates*, edited by D. Duvall, D. Müller-Schwarze, and R. M. Silverstein, 243–260. New York: Plenum.

Jansen, D. W., and R. C. Foehring. 1983. The Mechanism of Venom Secretion from Duvernoy's Gland of the Snake *Thamnophis sirtalis*. *Journal of Morphology* 175: 271–277.

Mason, R. T., and D. Crews. 1986. Pheromone Mimicry in Garter Snakes. In *Chemical Signals in Vertebrates*, edited by D. Duvall, D. Müller-Schwarze, and R. M. Silverstein, 278–283. New York: Plenum.

Schwenk, K. 1994. Why Snakes Have Forked Tongues. *Science* 263: 1573–1576.

Tyning, T. F. 1990. *A Guide to Amphibians and Reptiles*. Stokes Nature Guides. Boston: Little, Brown.

Vonstille, W. T., and W. T. Stille III. 1994. Electrostatic Sense in Rattlesnakes. *Nature* 370: 184–185.

Weldon, P. J., and D. B. Fagre. 1989. Responses by Canids to Scent Gland Secretions of the Western Diamondback Rattlesnake *(Crotalus atrox)*. *Journal of Chemical Ecology* 15: 1589–1604.

Plant Defenses

Anaya, A. L., B. E. Hernandez-Bautista, M. Jimenez-Estrada, and L. Velasco-Ibarra. 1992. Phenylacetic Acid as a Phytotoxic Compound of Corn Pollen. *Journal of Chemical Ecology* 18: 897–905.

Bergstrom, G., G. Birgersson, I. Groth, and L. A. Nilsson. 1992. Floral Fragrance Disparity Betwen Three Taxa of Lady's Slipper, *Cypripedium calceolus*, Orchidae. *Phytochemistry* 31: 2315–2319.

Bucyanayandi, J. D., J. M. Bergeron, and H. Menard. 1990. Preference of Meadow Voles *(Microtus pennsylvanicus)* for Conifer Seedlings, Chemical Components and Nutritional Quality of Bark of Damaged and Undamaged Trees. *Journal of Chemical Ecology* 16: 2569–2580.

Clausen, T. P., P. B. Reichardt, J. P. Bryant, R. A. Werner, K. Post, and K. Frisby. 1989. Chemical Model for Short-term Induction in Quaking Aspen *(Populus tremuloides)* Foliage Against Herbivores. *Journal of Chemical Ecology* 15: 2335– 2346.

English, S., W. Greenaway, and F. R. Whatley. 1991. Bud Exudate Composition of *Populus tremuloides. Canadian Journal of Botany* 69: 2291–2295.

Fackelmann, K. A. 1993. Food, Drug, or Poison? Cultivating a Taste for Toxic Plants. *Science News* 143: 312–314.

Gibbons, E. 1966. *Stalking the Healthful Herbs.* New York: David McKay.

Greenaway, W., J. May, T. Scaysbrook, and F. R. Whatley. 1992. Compositions of Bud and Leaf Exudates of Some *Populus* Species Compared. *Zeitschrift fuer naturforschung:* section C, *Biosciences* 47(5–6): 329–334.

Harris, R. W. 1992. *Arboriculture: Integrated Management of Landscape Trees, Shrubs, and Vines.* Englewood Cliffs, N.J.: Prentice-Hall.

Herman, A. E., and C. T. Robbins. 1993. Specificity of Tannin-Binding Salivary Proteins Relative to Diet Selection by Mammals. *Canadian Journal of Zoology* 71: 628–633.

Jakubas, W. J., and G. W. Gullion. 1990. Coniferyl Benzoate in Quaking Aspen: A Ruffed Grouse Feeding Deterrent. *Journal of Chemical Ecology* 16: 1077–1088.

Jakubas, W. J., and J. R. Mason. 1991. Role of Avian Trigeminal Sensory System in Detecting Coniferyl Benzoate, a Plant Allelochemical. *Journal of Chemical Ecology* 17: 2213–2222.

Lanza, J., M. A. Schmitt, and A. B. Awad. 1992. Comparative Chemistry of Elaiosomes of Three Species of *Trillium. Journal of Chemical Ecology* 18: 209–221.

Lindroth, R. L., and J. D. C. Hemming. 1990. Responses of the Gypsy Moth (Lepidoptera, Lymantriidae) to Tremulacin, and Aspen Phenolic Glycoside. *Environmental Entomology* 19: 842– 847.

McCourt, R. 1991. Some Like It Hot. *Discover,* August, 49–52.

Smith, B., P. D. Forman, and A. E. Boyd. 1989. Spatial Patterns of Seed Dispersal and Predation of Two Myrmecochorous Forest Herbs. *Ecology* 70: 1649–1656.

Yoon, C. K. 1992. Nibbled Plants Don't Just Sit There: They Launch Active Attacks. *New York Times,* June 23, section C.

Zhu, H., and A. U. Mallik. 1994. Interactions Between *Kalmia* and Black Spruce: Isolation and Identification of Allelopathic Compounds. *Journal of Chemical Ecology* 20: 407–421.

Fireflies

Abrahams, M. V., and L. D. Townsend. 1993. Bioluminescence in Dinoflagellates: A Test of the Burglar Alarm Hypothesis. *Ecology* 74: 258–260.

Borror, D. J., D. M. Delong, and C. A. Triplehorn. 1976. *An Introduction to the Study of Insects.* New York: Holt, Rinehart and Winston.

Carlson, A. D., and J. Copeland. 1988. Flash Competition in Male *Photinus macdermotti* Fireflies. *Behavioral Ecology and Sociobiology* 22: 271–276.

Milne, L., and M. Milne. 1980. *The Audubon Society Field Guide to North American Insects and Spiders.* New York: Alfred A. Knopf.

Stokes, D. W. 1983. *A Guide to Observing Insect Lives.* Stokes Nature Guides. Boston: Little, Brown.

Wing, S. R. 1991. Timing of *Photinus collustrans* Reproductive Activity: Finding a Mate in Time (Coleoptera: Lampyridae). *Coleopterists Bulletin* 45: 57–74.

Zimmer, C. 1994. Fireflies in Lockstep. *Discover,* June, 30–31.

Beavers

Müller-Schwarze, D. 1992. Beaver Waterworks. *Natural History,* May, 52–53.

Müller-Schwarze, D., L. Morehouse, R. Corradi, C.-H. Zhao, and R. M. Silverstein. 1989. Odor Images: Responses of Beaver to Castoreum Fractions. In *Chemical Signals in Vertebrates,* edited by D. Duvall, D. Müller-Schwarze, and R. M. Silverstein, 561–570. New York: Plenum.

Naiman, R. J., G. Pinay, C. A. Johnston, and J. Pastor. 1994. Beaver Influences on the Long-term Biogeochemical Characteristics of Boreal Forest Drainage Networks. *Ecology* 75: 905–921.

Novak, M. 1987. Beaver. In *Wild Furbearer Management and Conservation in North America,* edited by M. Novak, J. A. Baker, M. E. Obbard, and B. Malloch, chapter 25. Ministry of Natural Resources, Ontario.

Rue, L. R. 1964. *The World of the Beaver*. Philadelphia: J. B. Lippincott.

Welsh, G., and D. Müller-Schwarze. 1989. Experimental Habitat Scenting Inhibits Colonization by Beaver, *Castor canadensis*. *Journal of Chemical Ecology* 15: 887–893.

Mosses and Lichens

Brock, T. D. 1966. *Principles of Microbial Ecology*. Englewood Cliffs, N.J.: Prentice-Hall.

Bruemmer, F. 1991. In Praise of the Lowly Lichen. *International Wildlife* 21: 30–33.

Hale, M. E. 1967. *The Biology of Lichens*. New York: American Elsevier.

Rodbell, D. T. 1992. Lichenometric and Radiocarbon Dating of Holocene Glaciation, Cordillera Blanca, Peru. *The Holocene* 2: 19–29.

Thomas, R. J., A. J. Grethlein, C. M. Perou, and D. C. Scheirer. 1990. Translocation in *Polytrichum commune*, Bryophyta, part 3: Loading of Sugars in Source Leaves. *American Journal of Botany* 77: 1574–1581.

Bogs

Johnson, C. W. 1985. *Bogs of the Northeast*. Hanover, N.H.: University Press of New England.

Pepi, D. 1985. *Thoreau's Method: a Handbook for Nature Study*. Englewood Cliffs, N.J.: Prentice–Hall.

Prankevicius, A. B., and D. M. Cameron. 1991. Bacterial Dinitrogen Fixation in the Leaf of the Northern Pitcher Plant, *Sarracenia purpurea*. *Canadian Journal of Botany* 69: 2296–2298.

Rydin, H., and R. S. Clymo. 1989. Transport of Carbon and Phosphorus Compounds about *Sphagnum*. *Proceedings of the Royal Society of London:B, Biological Sciences* 237: 63–84.

Santelmann, M. V. 1992. Cellulose Mass Loss in Ombrotrophic Bogs of Northeastern North America. *Canadian Journal of Botany* 70: 2378–2383.

Schulze, W., and E. D. Schulze. 1990. Insect Capture and Growth of the Insectivorous *Drosera rotundifolia*. *Oecologia* 82: 427–429.

Stewart, C. N., Jr., and E. T. Nilson. 1992. *Drosera rotundifolia* Growth and Nutrition in a Natural Population, with Special Reference to the Significance of Insectivory. *Canadian Journal of Botany* 70: 1409–1416.

What Should You Do When You See a Bear?

Herrero, S. 1985. *Bear Attacks: Their Causes and Avoidance.* New York: Lyons and Burford.

Oftedal, O. T., G. L. Alt, E. M. Widdowson, and M. R. Jakubasz. 1993. Nutrition and Growth of Suckling Black Bears *(Ursus americanus)* During Their Mothers' Winter Fast. *British Journal of Nutrition* 70: 59–79.

Rogers, L. L., G. A. Wilker, and S. S. Scott. 1991. Reactions of Black Bears to Human Menstrual Odors. *Journal of Wildlife Management* 55: 632–634.

Schooley, R. L., C. R. McLaughlin, G. J. Matula, and W. B. Krohn. 1994. Denning Chronology of Female Black Bears: Effects of Food, Weather, and Reproduction. *Ecological Monographs* 75: 466–477.

Whitaker, J. O. 1980. *The Audubon Society Field Guide to North American Mammals.* New York: Alfred A. Knopf.

Young, S. 1985. Why Hibernating Bears Don't Starve. *New Scientist* 105: 17.

Conifers

Anderson, J. V., B. I. Chevone, and J. L. Hess. 1992. Seasonal Variation in the Antioxidant System of Eastern White Pine Needles. *Plant Physiology* 98: 501–508.

Halfpenny, J. C., and R. D. Ozanne. 1989. *Winter: An Ecological Handbook.* Boulder, Colo.: Johnson.

Harlow, W. M., E. S. Harrar, J. W. Hardin, and F. M. White. 1991. *Textbook of Dendrology.* New York: McGraw-Hill.

Little, E. L. 1980. *The Audubon Society Field Guide to North American Trees (Eastern Region).* New York: Alfred A. Knopf.

McLaughlin, S. B., M. G. Tjoelker, and W. K. Roy. 1993. Acid Deposition Alters Red Spruce Physiology: Laboratory Studies Support Field Observations. *Canadian Journal of Forest Research* 23: 380–386.

Percy, K. E., R. Jagels, S. Marden, C. K. McLaughlin, and J. Carlisle. 1993. Quantity, Chemistry, and Wettability of Epicuticular Waxes on Needles of Red Spruce along a Fog- Acidity Gradient. *Canadian Journal of Forest Research* 23: 1472–1479.

Zimmerman, M. H., and C. L. Brown. 1971. *Trees: Structure and Function*. New York: Springer-Verlag.

Princess Pines

Andrews, H. N. 1970. *Ancient Plants and the World They Lived In*. Ithaca, N.Y.: Comstock Publishing, Cornell Univ. Press.

Ayer, W. A., L. M. Browne, A. W. Elgersma, and P. P. Singer. 1990. Identification of Some L-numbered *Lycopodium* Alkaloids. *Canadian Journal of Chemistry* 68: 1300–1304.

Burns, G. W. 1974. *The Plant Kingdom*. New York: Macmillan.

Duman, J. G., and T. M. Olsen. 1993. Thermal Hysteresis Protein Activity in Bacteria, Fungi, and Phylogenetically Diverse Plants. *Cryobiology* 30: 322–328.

Headley, A. D., T. V. Callaghan, and J. A. Lee. 1988. Water Uptake and Movement in Clonal Plants *Lycopodium annotinum* L. and *Diaphasiastrum complanatum* L. Holub. *New Phytologist* 110: 497–502.

Soltis, P. S., and D. E. Soltis. 1988. Estimated Rates of Intragametophytic Selfing in Lycopods. *American Journal of Botany* 75: 248–256.

Woolf, G. M., L. M. Petrovic, S. E. Rojter, S. Wainwright, F. G. Villamil, W. M. Katkov, P. Michieletti, I. R. Wanless, F. R. Stermitz, J. J. Beck, and J. M. Vierling. 1994. Acute Hepatitis Associated with the Chinese Herbal Product Jin Bu Huan. *Annals of Internal Medicine* 121: 729–735.

Winter Woods

Harlow, W. M., E. S. Harrar, J. W. Hardin, and F. M. White. 1991. *Textbook of Dendrology*. New York: McGraw-Hill.

Palmer, E. L. 1949. *Fieldbook of Natural History.* New York: McGraw-Hill.

Whitaker, J. O. 1980. *The Audubon Society Field Guide to North American Mammals.* New York: Alfred A. Knopf.

Northern Lights

Akasofu, S. 1989. The Dynamic Aurora. *Scientific American* 260: 90–97.

Crooker, N. 1994. Replacing the Solar Flare Myth. *Nature* 367: 595–596.

Gosling, J. 1993. The Solar Flare Myth. *Journal of Geophysical Research* 98: 18937–18949.

Johnstone, A. 1984. Aurora Seen in Daylight. *Nature* 311: 413.

Keller, W. E. 1986. Why the Northern Lights Have Pleats. *Environment* 28: 23.

Livesay, R. J. 1989. A Jam-jar Magnetometer as "Aurora Detector." *Sky and Telescope* 78: 426–432.

Lockwood, M., and A. Coates. 1992. When the Solar Wind Blows. *New Scientist* 133: 25–28.

Saunders, M. 1992. Wonders of the Sky. *Nature* 360: 428.

Bird Feeder Biology

Barkan, C. P. L. 1990. A Field Test of Risk-sensitive Foraging in Black-capped Chickadees *(Parus atricapillus). Ecology* 71: 391–400.

Becker, C. D. 1993. Environmental Cues of Estrus in the North American Red Squirrel, *Tamiasciurus hudsonicus* Bangs. *Canadian Journal of Zoology* 71: 1326–1333.

Brittingham, M. C., and S. A. Temple. 1992. Use of Winter Bird Feeders by Black-capped Chickadees. *Journal of Wildlife Management* 56: 103–110.

Heinrich, B. 1993. Kinglets' Realm of Cold. *Natural History,* February, 5–9.

Hitchcock, C. L., and D. F. Sherry. 1990. Long-term Memory for Cache Sites in the Black-capped Chickadee. *Animal Behavior* 40: 701–712.

Klenner, W., and C. J. Krebs. 1991. Red Squirrel Population Dynamics I. The Effect of Supplemental Food on Demography. *Journal of Animal Ecology* 60: 961–978.

Popp, J. W., M. S. Ficken, and C. M. Weise. 1990. How Are Agonistic Encounters among Black-capped Chickadees Resolved? *Animal Behavior* 39: 980–986.

Price, K. 1992. Territorial Bequeathal by Red Squirrel Mothers: A Dynamic Model. *Bulletin of Mathematical Biology* 54: 335–354.

Smith, S. M. 1991. *The Black-capped Chickadee.* Ithaca, N.Y.: Cornell Univ. Press.

Stokes, D. W. 1979. *A Guide to Bird Behavior.* Boston: Little, Brown.

Sullivan, B. D. 1991. Additional Vertebrate Prey Items of the Red Squirrel, *Tamiasciurus hudsonicus. Canadian Field Naturalist* 105: 398–399.

Snowfleas

Bauer, T., and M. Pfeiffer. 1991. 'Shooting' Springtails with a Sticky Rod: The Flexible Hunting Behaviour of *Stenus comma* (Coleoptera; Staphylinidae) and the Counter-strategies of Its Prey. *Animal Behavior* 41: 819–828.

Borror, D. J., D. M. Delong, and C. A. Triplehorn. 1976. *An Introduction to the Study of Insects.* New York: Holt, Rinehart and Winston.

Lyford, W. H. 1974. Overland Migration of Collembola (*Hypogastrura nivicola* Fitch) Colonies. *American Midland Naturalist* 94: 205–209.

Milne, L., and M. Milne. 1980. *The Audubon Society Field Guide to North American Insects and Spiders.* New York: Alfred A. Knopf.

Whittaker, J. B. 1981. Feeding of *Onychiurus subtenuis* (Collembola) at Snow Melt in Aspen Litter in the Canadian Rocky Mountains. *Oikos* 36: 203–206.

Maple Sap

Ehrlich, P. R., D. S. Dobkin, and D. Wheye. 1988. *The Birder's Handbook.* New York: Simon and Schuster.

Gibbons, E. 1970. *Stalking the Wild Asparagus.* New York: David McKay.

Heinrich, B. 1992. Maple Sugaring by Red Squirrels. *Journal of Mammalogy* 73: 51–54.

Morselli, M. F., and M. L. Whalen. 1991. Aseptic Tapping of Sugar Maple *(Acer saccharum)* Results in Light Color Grade Syrup. *Canadian Journal of Forest Research* 21: 999–1005.

Zimmerman, M. H., and C. L. Brown. 1971. *Trees: Structure and Function.* New York: Springer-Verlag.

Index

245